Teacher's Handbook

Sue Atkinson

Sharon Harrison

Laurie Rousham

CAMBRIDGE
UNIVERSITY PRESS

PUBLISHED BY THE PRESS SYNDICATE OF THE UNIVERSITY OF CAMBRIDGE
The Pitt Building, Cambridge CB2 1RP, United Kingdom

CAMBRIDGE UNIVERSITY PRESS
The Edinburgh Building, Cambridge CB2 2RU, United Kingdom
40 West 20th Street, New York, NY 10011-4211, USA
10 Stamford Road, Oakleigh, Melbourne 3166, Australia

© Cambridge University Press 1998

First published 1998

Printed in the United Kingdom at the University Press, Cambridge

Set in Swift and Frutiger

A catalogue record for this book is available from the British Library

ISBN 0 521 63428 8 paperback

Notice to teachers
The photocopy masters on pages 128–36 of this publication may be photocopied free of charge for
classroom use within the school or institution which purchases the publication. Worksheets and
photocopies of them remain in the copyright of Cambridge University Press and such photocopies
may not be distributed or used in any way outside the purchasing institution. Written permission is
necessary if you wish to store the material electronically.

Contents

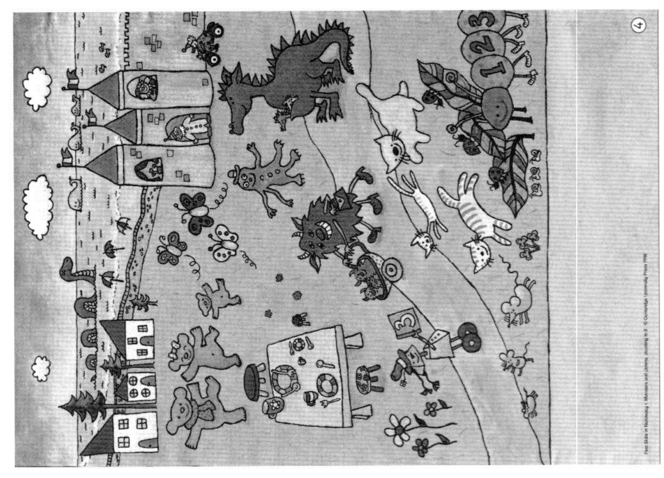
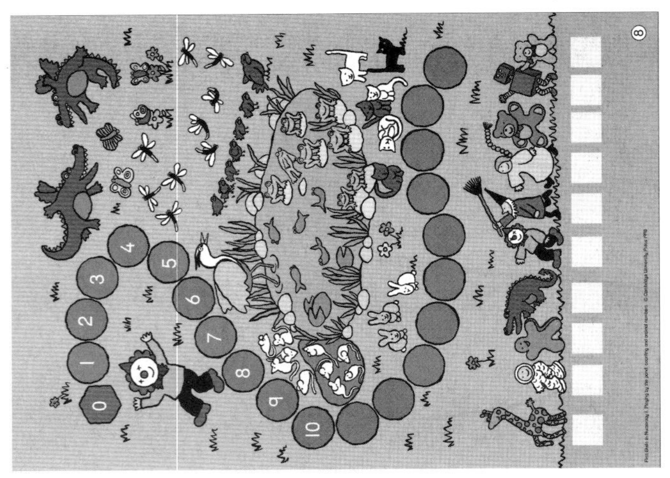
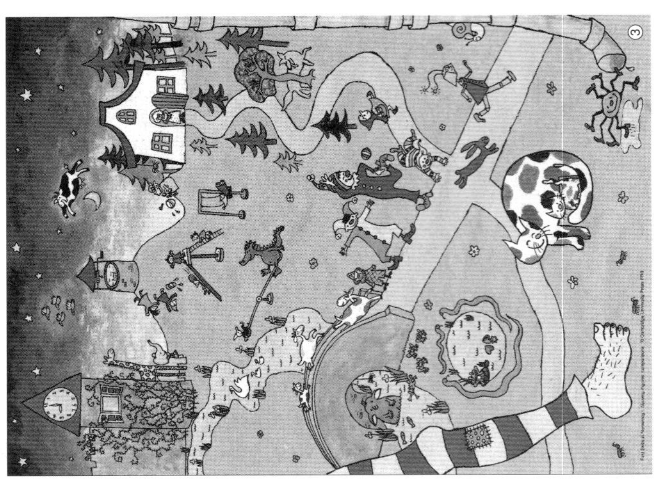
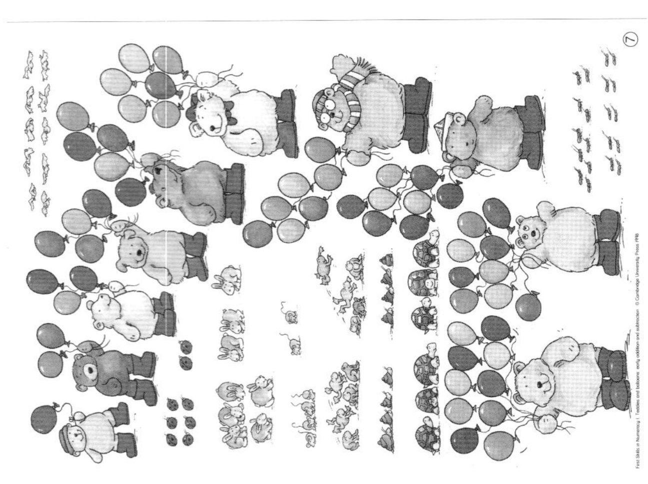
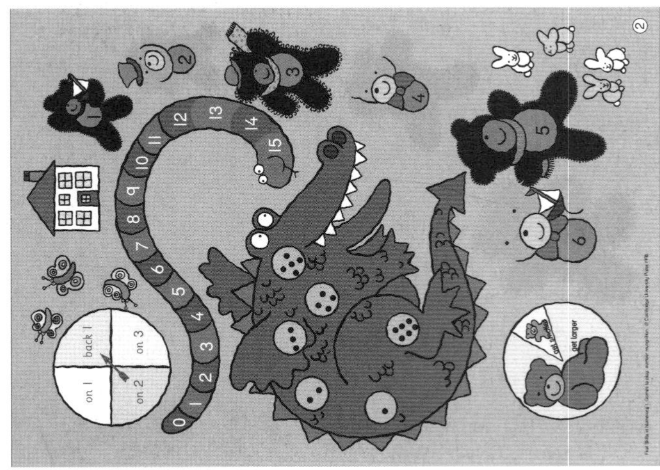
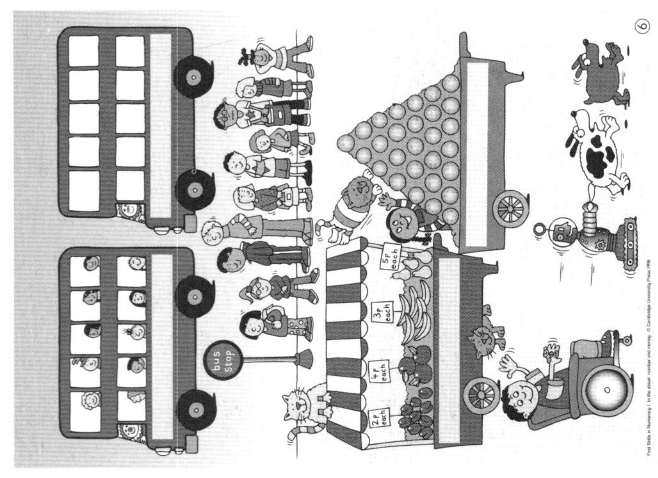

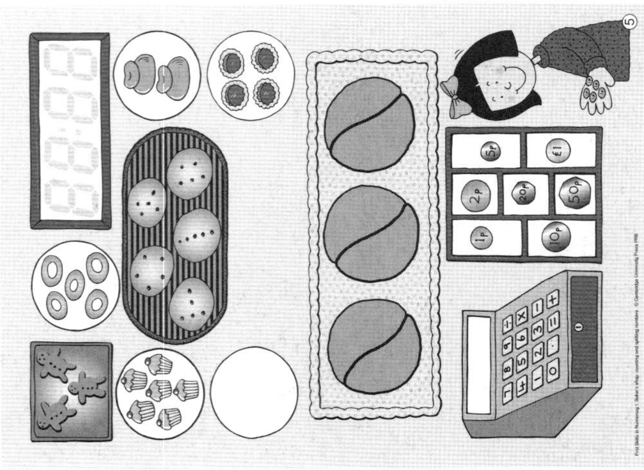

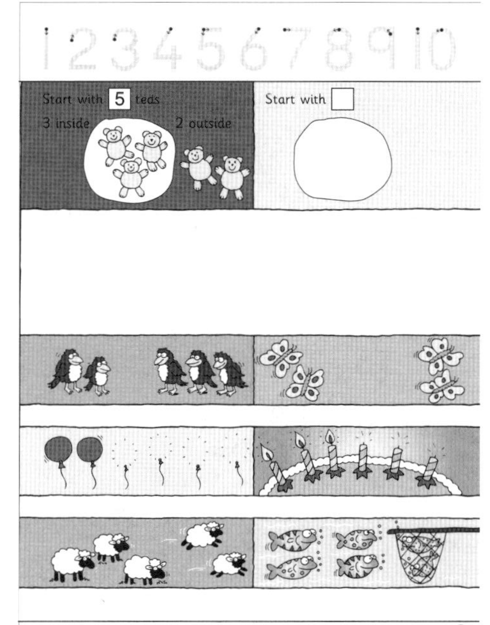
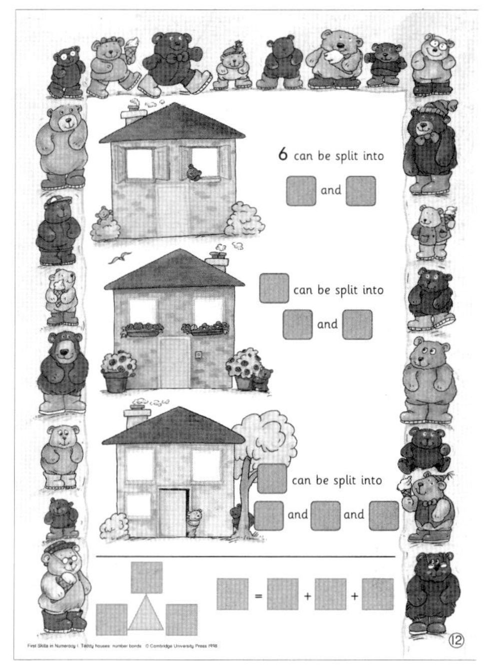
123456789 10
Start with 5 teds
3 inside
2 outside
Start with

6 can be split into   and
can be split into   and
can be split into   and   and
=   +   +

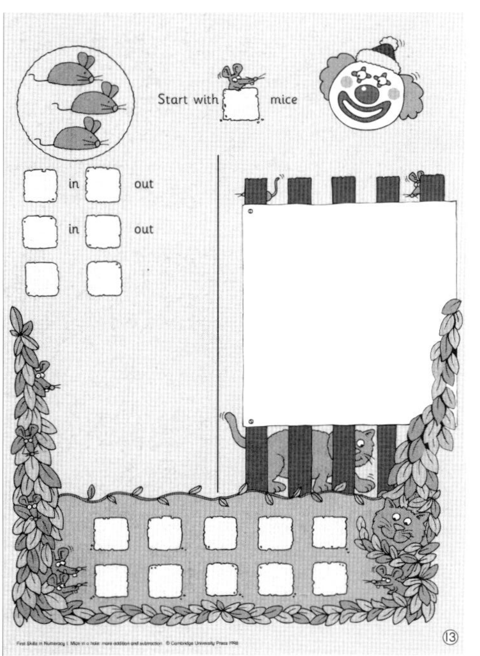
Start with   mice
in   out
in   out

5   6   7
8   9   10
You can write your sums here

1 2 3 4 5 6 7 8 9 10
11 12 13 14 15 16 17 18 19 20
21 22 23 24 25 26 27 28 29 30
31 32 33 34 35 36 37 38 39 40
41 42 43 44 45 46 47 48 49 50
51 52 53 54 55 56 57 58 59 60
61 62 63 64 65 66 67 68 69 70
71 72 73 74 75 76 77 78 79 80
81 82 83 84 85 86 87 88 89 90
91 92 93 94 95 96 97 98 99 100

What is First Skills in Numeracy?

First Skills in Numeracy

- helps you teach numeracy more interactively
- develops children's own mental strategies
- enables you to spend more time teaching and less time organising

First Skills in Numeracy draws on the work of the National Numeracy Project and others. It is also in line with the aims and objectives set out in *Desirable Outcomes for Children's Learning on Entering Compulsory Education* (SCAA, 1996). It shares the National Numeracy Project's emphasis on whole class teaching, oral and mental work and direct communication with pupils.

The National Numeracy Project also recommends that numeracy lessons have a three-part structure. These stages are a whole class starter, a main activity or group time followed by a plenary session when the major concepts of the lesson are reviewed. The lesson plans in FSIN follow this pattern.

The materials in First Skills in Numeracy

FSIN 1	FSIN 2	FSIN 3
16 Interactive Pictures	16 Interactive Pictures	16 Interactive Pictures
Teacher's handbook	Teacher's handbook	Teacher's handbook
Practice book 1	Practice book 1	Practice book 1
Practice book 2	Practice book 2	Practice book 2
Practice book 3	Practice book 3	Practice book 3 – SATs practice
Handwriting book		
Photocopiable version of workbooks	Photocopiable version of workbooks	Photocopiable version of workbooks

Interactive Pictures

- form the starting point and context for the lessons and other activities in the teacher's book
- act as prompts for class and group discussion
- contain some smaller detail for children working in groups

There are 16 large colour pictures for each year. They are easy to use: simply clip to an easel or fix to a wall for starting a whole class lesson or lay flat for group work.

Their write-on wipe-off surface is suitable for use with washable spirit pens.

They may be stored flat or kept rolled in their tube.

Teacher's book

FSIN 1 teacher's handbook contains:

1) Information to help you get started

This includes detailed instructions on how to get the most from your numeracy lesson.

2) Activity bank

Select from these activities for your First Skills lessons or to supplement your own maths scheme, to fill odd five minutes during the day or to keep number work ticking over while concentrating on a shape and space topic.

- Mental maths and circle games:
  This section shows you how to organise mental maths in an early years classroom and so help children begin to develop their calculating skills. The activities are arranged by type:
  mental maths 1 – examples of questions that can be used at any time to develop children's numeracy

mental maths 2 – activities which develop children's visualisations in maths
mental maths 3 – practical mental maths activities that we have called circle games

- Structured play
  This section is appropriate for both nursery and reception. It outlines the intended learning outcomes, lists the equipment needed and gives suggestions for other related activities.
- Maths table
  Ideas are given for setting up a maths area within the classroom. This can be used as a focus for both structured and free play activities that children can do reasonably independently. Examples of activities and learning objectives are given.

3) Lesson plans

- Nursery lesson plans
  These are short starting points based on the Interactive Pictures (IP).
  They would also be useful activities to use with very young reception children at the beginning of the year.
- Reception lesson plans
  These follow the three-fold structure of 'plan, do, review' or 'introduction, group time and review' preferred by the National Numeracy Project.
  They are each linked to an Interactive Picture and this provides the context for the lesson.
  The different parts of these lessons are explained in the blueprint for a lesson plan (see pages 10–11).
  You will want to repeat many of the lessons and activities several times and develop them further. Ideas are given of ways to do this.

4) Photocopiable material

- Resource sheets
  There are nine of these to be used by children working at the maths table or as assessment activities.
  You can differentiate the activities for different children by using 'white out', or by writing in numbers appropriate for particular abilities. Adding pictures with rubber stamps creates variety and makes them a very versatile resource.

5) Reference pages

- The planning grids will enable you to find activities that will help you to teach a particular learning objective. They draw heavily on the Framework of Objectives of the National Numeracy Project and the Desirable Outcomes as published in *Desirable Outcomes for Children's Learning on Entering Compulsory Education*.

FSIN 1 Practice books

These are also available in a photocopiable format.

Although formal recording is not essential in the early stages of education, encouraging children to draw and write what they plan to do and what they find out gives you something to focus on during review sessions and provides you with evidence of children's learning.

1) Practice books

These are intended to be used for more formal recording in reception classes.

The games on each cover can be taught to the whole class using IP 2.

Children can work on the pages after certain whole class starter activities and circle games. Some pages may be used for assessment after an activity and others provide practice for consolidation of key skills.

Ideas and concepts are developed from page to page in increasing levels of difficulty.

Some children will be able to complete most pages independently. However, working alongside them as they do so will enable you to assess their mathematical language, and support and extend their learning.

Some children find formal recording very hard, but this does not mean that they do not understand the mathematical concepts.

2) Handwriting book

This can be used alongside the practice books, or completed first. It gives practice in the writing of numbers and, towards the end of the book, some counting work to 10.

First skills 1 – Getting started

Raising standards of numeracy in the early years

Children can do much more than was previously thought. We can observe two and three year olds sorting and matching successfully and so, although sorting and matching activities are included in First Skills, ideas have also been given to show how these skills may be developed to reflect the needs of the numeracy curriculum, giving children the opportunity to achieve their full potential.

We need to expect more from children and give them experiences with numbers larger than was previously thought appropriate.

It is important for children to continue to handle objects and use their fingers, bricks etc. for counting while they gradually develop an ability to manipulate numbers mentally.

You will find that, in order to provide a firm foundation for later learning, a great emphasis is placed in First Skills on developing a wide mathematical vocabulary and on giving children repeated experiences of working with number lines.

Raising the profile of maths in the early years

- Use maths displays as stimuli for work and discussion during the numeracy hour.
- Display numbers and charts, eg the coffee making rota for the staff, the number of children allowed to play in different areas, numbers on pegs and drawers for tidying up, eg '4 hats', '10 painting aprons'.
- Give status to mathematical play by sitting with children to discuss what they are doing.
- Give thoughtful and constructive praise, extending 'what a good fire-engine' to 'your fire-engine is longer than Jack's and has more bricks'.
- Make the most of all opportunities to develop children's understanding, eg when asking children how many Polydrons they used, remind them to use only one number name for each Polydron and count alongside them.
- At circle time, spend time on teaching children how to explain ideas carefully and how to listen to others; good quality mathematical discussion is crucial to learning new ideas.
- Develop the habit of asking children what they have learnt during the day. You can list this on the board.

- You need to count something with children every day in both nursery and reception.

Planning

Your school's scheme of work for maths will specify the concepts that are to be covered in your class. The planning grids at the back of this teacher's book draw heavily on the work of the National Numeracy Project and the *Desirable Outcomes* and you will be able to to match the learning objectives with those in your existing teaching plans.

The nursery and reception lesson plans are each linked to an Interactive Picture and follow the order of the IPs. It is not intended that you start teaching the lessons that relate to IP 1 and then move on through the other IPs in order. Instead, use the planning grids to find the lessons that focus on the learning objective you are trying to achieve. The facsimiles at the front of this book give the overall learning objectives associated with each page. (In FSIN 1, most of the pictures provide opportunities to work on several aspects of mathematics over the two years.)

All the lessons and other activities described in this teacher's book can be found on the planning grids. In addition, the activities in the activity bank sections are arranged by learning objective, allowing you to find the one that suits your purposes easily.

The grids also include the important mathematical concepts that children will need to cover in nursery and early reception before they move on to the more formal numeracy curriculum.

Using the reception lesson plans

These are arranged across two double-spreads. You will find a detailed lesson plan on the first double-spread.

The blueprint on pages 10–11 shows what happens in each part of the lesson.

Over the page there are further lesson plans. These usually have different learning objectives and use different aspects of the IP from those used in the first lesson plan. They might also be at a different level of difficulty and so you would not teach them at the same time in the year. These lesson plans are not as detailed as the first and some are given in outline only.

Once you become familiar with the materials you will find it easy to make up your own lesson plans using activities from different sections of the teacher's book. In fact it is not necessary to do everything in a lesson at one go as you may feel that some activities are not appropriate for the children in your class.

Questions for the teacher to ask are in italics.

Equipment needed

The following is a list of items that you would find it useful to have when using First Skills.

1. Tactile number cards

Number cards made of felt, or glue and sand, or corrugated card can be used in a variety of ways in nursery and reception.

Make cards with a coloured strip at the bottom so that children know which way up the number is. If this strip has a tactile quality, eg felt, the children will be able to feel numbers in a feely bag and know which way up to hold them.

Paint on white glue in the shape of the numbers and sprinkle this with dry sand. When the glue has dried, shake off the excess sand.

2. Dotty cards

Make dotty cards with stickers (or with sand and glue as above),

- in dice patterns 1–10
- with random dots
- with regular rows of dots especially in pairs
- for estimating numbers, eg almost 20 (19 and 21), near 5 (4 and 6),
- vary the 'dots': different-sized dots; pictures (use all the same picture on one card); represent all the numbers to 20, but with several different arrangements of dots, eg show 5 as a dice pattern, in a well spaced row, bunched close together in a row, as 3 small stars and 2 large ones, as 3 cars together and 2 more.

3. Several packs of number cards

- for a small group, enough 0–5 for each child, a few packs 0–10
- for a whole class, 0 to at least 20,
- number word cards ('one', 'two' etc.)

4. Dice and spinners (available from Tarquin 01379 384 218).

5. A variety of number lines and tracks.

6. Rubber stamps and felt pens.

7. A book of number rhymes, eg *Seven Dizzy Dragons* Cambridge University Press 1997

Lesson planning

There are many ways to structure a maths lesson and the format you choose will depend on different things, such as the age of the children in your class, the policy of your school.

One example of a structure for a lesson is shown on pp 10–11. This type of lesson could be used with the whole class or with half the class while the other children do something other than maths.

Whole class starter

1. Seat the children either on the carpet or at their tables where you can see one another easily.

Start with a mental warm-up, going over recent work and known number facts.

2. Tell children what they will be learning in this lesson. Use a picture, the board or objects as a focus for children's attention.

3. Introduce new strategies and ideas through interaction with the children:

Demonstrate new ideas, eg counting in 10s.

Explain why a particular method works.

Question children to find out what they have understood so far. Use open rather than closed questions to encourage children to reflect on and talk about their ways of working problems out, eg 'What did you do to find your answer?'

Main teaching activity / Group time

1. Splitting the class into four ability groups will enable you to work intensively with half the class one day. The following day you can rotate the groups so that everyone tackles all the activities (differentiated for their abilities) or you might choose a new task for a group that needs support or extension work.

2. Guide your focus group through new ideas or extend the starter. This is a time for in-depth assessment.

Model mathematical language and encourage children to be flexible in the mental strategies they use when approaching a calculation.

Identify and rectify any misconceptions held by children as they crop up.

3. The remainder of the class use this time for practice and reinforcement of the ideas from the whole class starter. This could be a practical or written activity, but in either case it should be clear to them that they will be presenting what they have done as part of the discussion at review time at the end of the lesson. Being aware of this can help children to keep on task during this part of the lesson.

Lesson plan blueprint

Learning objectives
These are the learning objectives that may be covered by lessons linked to this picture. Each of the lessons below covers one or more of these objectives.

IP number and title
This title is also on each IP.

Lesson 1 Activity title
Although the lessons are numbered, they do not necessarily follow on from one another.

You need
These are the items you need for the whole class starter. You will need other things for the later group work.

Key words
This is the mathematical vocabulary that you need to introduce into your discussions with the children.

Initial assessments
During the whole class starter, use this list to note evidence of children's understandings. The information you gain can help you group children for the next stage of the lesson.

Level
As many tasks are open-ended, the levels given are broad. You can repeat the lessons, working at a different level each time you do so, to develop children's thinking.

Objectives
There is often more than one learning objective. You can choose to cover only one of these at a time, or all of them with more experienced children.

Whole class starter
The whole class starter often includes a circle game from the circle games section. See page 9 for information about this part of the lesson.

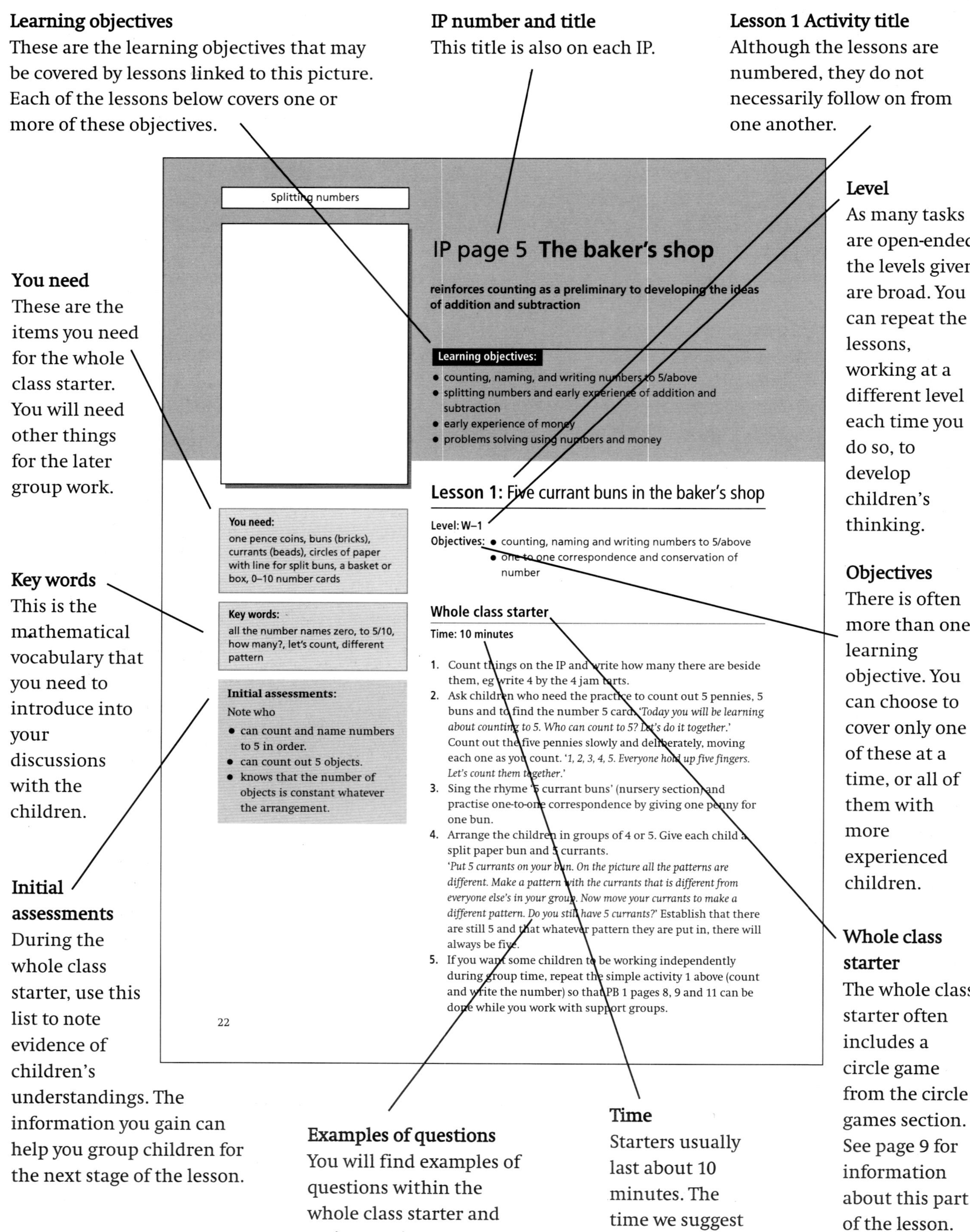

Examples of questions
You will find examples of questions within the whole class starter and review sections.

Time
Starters usually last about 10 minutes. The time we suggest is just a guide.

Group time

Usually, children in different groups will work on different activities, but the whole class could be doing the same activity, eg drawing what they did for a calculation, or all playing PB cover games.

See page 9 for more information about this part of the lesson.

Working more independently

Most children should be able to complete these activities with very little teacher support.

Support and extension activities are suggested where they are appropriate.

If children are going to be working in a PB, a facsimile of the relevant page will be shown.

Teacher focus

Usually we give an example here of what to do with average children, but sometimes you will want to have extension or support groups.

This is an opportunity for you to make assessments of individual children.

Whole class review

This whole class time is for children to show what they did if that is appropriate. You will not usually have time to let every child explain what they did.

See page 12 for more information about this part of the lesson.

Maths table

This is a list of the resources needed specifically for this lesson. You will find details of the equipment that you should always provide on pages 42–3 in the maths table section.

Alternative starters

Use these ideas, if you want to repeat this lesson so that children have the opportunity to do all the group time activities.

Key idea

Make this the focus of your discussion.

Time

Usually between 5 and 10 minutes.

Group time

Teacher focus

Time: 15 minutes

1. Count a variety of objects and write numbers appropriate to the children's experiences eg put out objects in 2s or 3s, or match objects to number cards, trace over tactile number cards.
2. Keep the IP on display. Ask individuals *'What can you see?'* to get a more detailed picture of the child's knowledge of numbers and counting. Note who can say the number words in the right order independently.

Working more independently

A. **Maths table**: Put out paper and interesting pens, number cards, a number track to 10, sorting objects and trays. *'Can you write the numbers 1, 2, 3, … Put that number of counters by each number.'*
B. *'Lay the table for the 3 bears/play people using cups, plates etc.'* Emphasise that each bear needs one cup, one plate etc to encourage one to one correspondence. *'How many plates? How many bears? Would there be enough plates if another bear came along? How many more would we need?'*
C. **Support**: *'Make playdough buns, one for everyone in the group/class, each one with 3 currants on top.'*
D. **Extension**: *'How many can you count up to?'*

Whole class review:

Time: 10 minutes

Key idea	Counting

1. Ask children to show their work. Help children to identify their threes. *'Can you count the currants on your bun for us?' 'How many plates did you need for the three bears?' 'Did you have more plates or more cups or the same number?'* Encourage children to ask one another questions. *'Who would like to ask Rik a question about his robots?'*
2. Practise writing numbers in the air once more and demonstrate how to use the tactile numbers again.
- Note who will need support with numbers to 3 and who can be extended beyond that.

Alternative starters:

1. *'Let's see if we can count up to 5 and count out 5 children. Who can see a 5 on the page?'* (Include all children in this work, even if they are still working with lower numbers.)
2. Read the Three bears / Billy goats gruff or make up 'Snow White and the 3 dwarves' or 'The 3 Dalmatians'!
3. Trace all the numbers 1–10 and recite up to 10. This will be by rote for some children but that is a necessary stage.

23

Review time
1. Encourage children to talk about their work and what they have learnt.

Ask questions yourself and invite other children to ask questions too. Children will learn how to communicate mathematically through such exchanges.

2. This part of the lesson is an opportunity for you to make informal assessments of children's progress, noting any misunderstandings or gaps in their knowledge.

3. It can also help you to plan follow-up work and set targets for individual children more effectively.

4. Round off the lesson by reflecting with the class on what has been learnt and what needs to be remembered for the future.

The transition from one part of the lesson to the next should be as smooth as possible and the pace of the lesson kept up.

Assessment
Both formal and informal assessment ideas are given throughout the text of the teacher's book.

Informal assessment
1. The Initial assessments section in each reception lesson plan gives you things to look out for at the start of the lesson to help you decide who needs help with the work in hand.

2. During group time you will be able to make assessments of the children working with you in your focus group.

3. The Whole class review suggests questions you can ask to assess who will need more work on the topic and who is ready to move on.

Formal assessment
The back page of PBs and PB 1 page 15 are used for specific assessments of maths content.

Quick ways into using First Skills for supply teachers and others
- Mental maths and circle games require very little preparation, organisation or equipment and they will give you insight into children's levels of understanding and experiences.
- The play and maths table sections take a little more organisation but can be set up when you want to settle children quickly into fairly teacher-independent tasks.
- The planning grids at the back will enable you to see where aspects of the numeracy curriculum are covered in First Skills.
- Each of the pupil books has a game on the cover. These can be taught to the whole class using IP 2 and played fairly independently.
- The lesson plans all start with whole class activities which will enable you to assess children's experiences and needs.

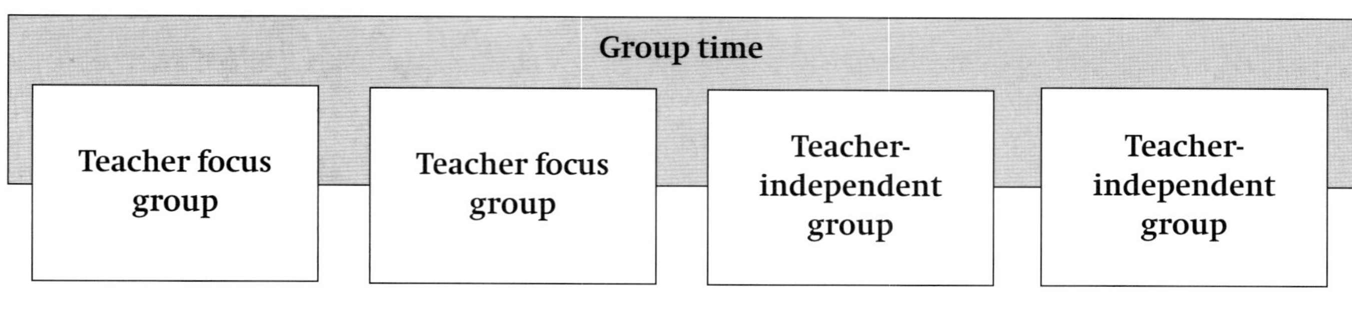

Activity bank

The development of mental methods is central to children's progression in numeracy. As teachers involved in the National Numeracy Project have found, giving time daily to oral and mental maths sessions leads to a raising of standards in numeracy.

There are many aspects to mental maths including understanding how the number system works, knowing number facts, using appropriate strategies when calculating and problem solving.

This section shows you how to introduce mental maths sessions into your classroom and gives you ideas for helping children to develop flexible strategies for calculating. It also includes activities for you to do with your class and questions to ask during mental maths sessions.

How to set up mental maths in your classroom

In order to make the most of the time that is spent on mental maths, it is necessary to establish a calm way of working so that each child feels safe, able to contribute and feels unpressurised with time to concentrate. You can do this by

- giving children plenty of time to work out answers and not allowing children to call out.
- having an agreed signal that children use when they want to speak.
- insisting that everyone else listens to other children.
- trying to ask children to answer only if you are fairly sure they know roughly what is expected of them. (Try not to pounce on day dreamers!)
- always starting with very easy questions so that everyone feels able to contribute.
- maintaining eye contact with the whole class and alerting children who are switching off that they need to pay attention.
- doing a whole class activity every few minutes to keep children alert.
- keeping your language as clear and as varied as possible.
- not always being the one with all the answers. Ask: *'Is she right?' 'What do you think about that?' 'Tell us what you are thinking.' 'Has anyone got a different answer?'*

Above all, make times together on the carpet positive and reassuring so that children will want to contribute and feel that they are good at maths.

- Be lavish with constructive praise to boost self esteem and confidence and give good feedback, eg *'Lisa, that's the first time I've seen you count back – well done.'*
- Respond to all contributions so that everyone feels valued.

Developing children's flexible methods of calculating

Strategies for calculating

There are many different strategies and ways of calculating. Below is a list of those you are likely to find children using in nursery and reception, followed by a list of more advanced methods.

The strategies and ways of calculating that you are likely to come across in nursery and reception are:

- counting each time from 1.
- counting on and back in 1s with counters/ fingers.
- counting on and back in 1s, mentally.

More advanced strategies that you may encounter are:

- using previous knowledge, eg 'I know that 2 and 2 is 4 so 2 and 3 is one more – 5.' (Laura age 5 working with her fingers)
- starting to see that they need not count from 1 every time (conservation of number, ie 'If that line of cubes was 3 a minute ago, it is probably still 3 so I do not need to count it again.')
- beginning to understand that with a calculation such as 1 add 6 more, they can start with the larger number, the 6 (without counting those counters/fingers again) and count on the smaller number.
- starting to see a quicker way than counting in 1s every time, eg counting in 2s.

(These strategies are not necessarily learnt by children in this order and not all children will have learnt to use these strategies by the time they are six.)

Teaching children the language of calculating

In order to develop children's existing strategies and move children on to using new ones, it is necessary for them to be able to talk about the methods they are currently using.

Your aim in nursery is to develop a language for talking about the practical counting and calculating that is being done with a wide range of equipment from cubes to number lines. It is also to use a range of different mathematical words within contexts (real and play) such as cooking, shopping and number rhymes.

Talking about what we actually do as we calculate is hard for anyone. At first, therefore, young children need to develop some sense of 'at homeness' with numbers and a familiarity with the language of calculating. With a calculation such as 1 and 4 more, children will work out the correct answer, perhaps on their fingers, but may not understand what is wanted when we ask 'How did you do that?' It can take some time for children to grasp what it is they have to do when we ask them to tell us what they did in their heads.

This work is extended in reception so that children begin to be able to talk about how they do calculations. We can help children to do this by giving them plenty of opportunities to share their own intuitive methods and supporting their language development.

How do I teach this?

The first lesson for IP 10 shows how to introduce children to the idea of talking about how they calculate. You will need to repeat the content of lesson 1 for IP 10 several times during mental maths time. Children will gradually begin to analyse and talk about what is going on in their heads.

1. Start with a simple calculation, eg *'2 teddies and 4 more teddies would make how many teddies altogether?'* Then let children talk about what they did, with you repeating and clarifying what they say.

2. Explain that there is more than one way to do some calculations. Give experiences of doing calculations in different ways in your mental maths sessions. At first these 'different ways' are likely to be the use of different equipment. This is a useful stage as it will help children become accustomed to talking about what they are doing.

For example, Matt might get out the cubes, Celia might jump along the number line and Beth might use her fingers.

Eventually, when children are talking confidently, ask more sophisticated questions, eg *'There are 5 speckled frogs on a stone. 2 of them jump in the pool. How many left?'* This could be worked out either by counting on or counting back on a number line, or by counting on fingers.

3. Sharing and comparing strategies is crucial. At review time, help confident children to explain what they did and use that to move other children on, eg *'Let's all do this calculation like Amy without counting our counters all the way from 1 again.'*

4. Very careful observation and listening is necessary throughout this process. It is likely that most children will count in 1s at first, usually counting from 1 each time, eg with 2 + 4, children will hold up those fingers, but count from the beginning each time even when you tell children to check what they did. Even if they have just counted the 2 fingers, they will count them again from 1.

How do I move children on?

1. Teach specific number facts, eg:
- easy number bonds such as 1 and 1 makes 2, 2 and 1 makes 3 (see 'Five little astronauts' *Seven Dizzy Dragons*, page 3).
- number bonds to 5 and then 10.
- doubling (and halving) numbers up to at least 5.

2. Teach specific skills, eg:
- emphasise that you don't need to start from 1 again each time you count: *'That was 4 counters last time we counted and we haven't moved any of them, so there must still be 4.'*
- demonstrate how it is quicker to reverse calculations where one number is larger than the other, eg 2 + 5: *'Change it round so that you hold the 5 in your head and just count on 2 on your fingers, like this: 6, 7.'*
- show children how to use what they know to work out other calculations, eg they could use known doubles to work out 2 sweets and 3 more sweets or 3 frogs and 4 more frogs.

3. Praise children whom you observe using more advanced strategies. Sit with confident children as they calculate and ask questions such as:
- *'How many were in that first set last time you counted? So do you really need to count them again?'*

- *'Is there a quicker way of doing that?'*
- *'Can you see another way to do that?'*
- *'Hold that number in your head.'* (instead of counting them all from 1 again)

Working with more confident children in this way may help them to move on to the next stages. You can then try to draw out their experiences in review time so that other children will learn from them.

4. As children share what they did, our response and tone of voice has an impact on how willing they will be to tell us what they are thinking. It helps to say:

- *'How did you do that?'*
- *'That was really great but I'm sure someone did it a different way.'*
- *'Can anyone think of a different way?'*
- *'Tell me how you did it, Charlie.'* Let Charlie explain, then try to say back to him what he said, *'So you did it by ...'*
- *'Did anyone else do it Charlie's way?'*

5. Aim, by the end of reception, for children to be
- moving beyond doing all their counting and calculating by counting in 1s. (Of course, at this early stage many children are going to be using counting in 1s as their usual method of calculating and we must praise their accuracy and build up their confidence with plenty of practical work with fingers and counters.)

- able to talk about what they are doing.
- able to use simple strategies to fit with the numbers they are using (eg able to use near doubles to calculate 5 + 6 but use counting on along a number line for 10 + 3).

It is not usually helpful to expect children to calculate quickly at this stage. Although some children's own methods are very slow, it may be some time before it is appropriate to suggest that they look for quicker methods.

Activities and questions to ask for mental maths

These are divided into three main sections.

Mental maths 1: questions and starting points for
- daily mental maths sessions and five minute fillers.

Mental maths 2: questions and starting points for
- helping children to visualise ideas in their minds.

Mental maths 3: circle games (practical mental maths activities)
- for whole class practical starter activities.

Mental Maths 1

The activities in this section are designed for use
- in a mental maths 'warm up'.
- in a self-contained oral maths session.
- on days when you are doing shape and space but you want to keep number ticking over.
- in a short session to see where children are, eg for use by a supply teacher.
- when you want to review concepts already covered.

These are a 'mixed bag' of questions to give an idea of the different ways in which we can phrase oral maths to give children experience of a wide range of vocabulary.

Counting

You need to count something with the children every day in both nursery and reception.
- Number rhymes and counting books provide a rich source of ideas for counting. Try to include rhymes each day as they work well. This is because in learning them by heart children learn the appropriate vocabulary at the same time and also because the numbers are in a context that they can understand.

Questions to ask:
- How many fingers am I holding up? Count them in your head. Now you hold up the same number of fingers.
- Who can say '1, 2, 3' like that all on their own? Let's say it together. Now let's count to 4.
- Stand up you three children. Let's give them a

toy each. How many toys? What if we lay out those toys in a long line? How many now?

- How many dolls am I holding? (Have them different sizes.) Let's count them, 1, 2, 3, 4. What if I put 2 down on the floor and sit 2 on my lap? Will there still be 4 dolls? What if I squeeze them all closely together?
- Let's count the birds on the mobile. (This gives experience in counting items out of reach.)
- Let's all count to 10. Now let's do it very loudly. Now let's do it very quietly and hold up a finger for each number as we say it.
- I'm going to show you a number on a card. If you know the number, whisper it very quietly and hold up that many fingers. What number is one more/less than that?
- Estimate how many dots on this card. Is it about 3 or about 10? Now let's count them.
- Estimate how many flowers there are in this vase. Now let's count them.

Harder counting

- I'm going to show you a dotty card. If you know how many dots there are on it, hold up a 'thinking thumb' and I will ask someone to come out and find the number card it goes with. (Use both randomly arranged dotty cards and those in dice pattern groups.)
- I'll say a number and you count back from that to zero.
- I'm going to drop these stones into the bucket (out of sight). Count the sounds you hear, silently in your head. How many?
- Count in 2s with me, 2, 4, 6, etc.
- Say one of the numbers that comes between 6 and 9.
- Let's count all the children in class today. How could we do it without counting anyone twice?

Mental calculation and ordering

- Tell me which number comes just after 5 / one before 7 on the number line.

- Let's muddle up the number cards on the carpet, then see if we can put them in order.
  Now they are all in order, close your eyes and I am going to take two of them and swap them over so they are out of order. Open your eyes and tell me which ones I moved.
- Pretend you are standing on 3 on the number line. Take a big step of 2 forwards. Where will you land? What if you now go back 4 steps?
- Can you do these in your head? 1 frog and 2 more frogs. 1 less than 4. 2 and 2. How many does 1 and 3 more come to?
- The answer is 3. Can you think what the question could be?
- You 5 children stand up. You 2 sit down now. How many still standing?
- I've got 4 bricks behind my back. Here is one of them. How many are still behind my back?
- Which is more, 3 or 5?
- Is 2 fewer than 3 or more than 3?
- Start with 4 balloons. Oh dear, 2 have popped. How many are left? How did you work that out?

Re-visiting previous work

- Who can remember that quick way to … we talked about yesterday?
  For example, after a lesson on reversing the numbers when adding, to put the larger one first, and then counting on: 'Who can remember the quick way to add two numbers like 2 and 7?'
- Do you remember last week when we counted to 20? Think about those numbers. Think about how 13 is written and write it in your mind. Now write it in the air.
- Remember last week when we were weighing. What happened to the balance scales when we put something heavy in it?

Mental Maths 2

The next section contains ideas for mental maths activities that mainly involve children learning to visualise. Some of these follow on from a practical activity as in the circle games and from lessons based on IPs. You would usually do these visualising activities alongside practical work.

Sorting and matching

■ identifying attributes

MM 2 I'm thinking about a bear **IP 2**
1. 'I'm thinking about a bear and it is holding a boat / medium-sized with a red bow tie / the smallest bear / has a blue hat on / has this number (hold up a card) on its front. Come and point to the bear I am thinking of.'
2. 'Shut your eyes and make a picture of a bear in your mind. Tell us about your bear.'
3. Move on to making groups or sets of bears. Choose 2 bears from IP 2 with a common attribute. 'This bear and this bear are in my set. Tell me about my set.' (dark coloured bears, bears holding a boat, bears with a hat/scarf on etc.)

Number

■ ordering numbers

MM 4 Number lines **IP 14**
1. 'Shut your eyes and pretend that you are writing numbers on the number line. Which number are you starting with?'
2. 'Now write the numbers in order from 0 and stop when you get to 3. Which numbers have you written?' 'Now go on writing but stop at 6. Tell me one of the numbers between 3 and 6.'
3. 'Tell me a number after 4 but before 10.'

■ visualising objects

MM 1 Bears and more bears
1. 'Think of three bears, a little one, a middle-sized one and a big bear. Now tell the middle-sized one and the little one to go away. Along come two more big bears. Put your big bears in a row, 1, 2, 3. Now make one of them stand a little way away from the others, so you have two bears and one more bear. One now goes right away.'

Continue adding bears and hiding them and putting them in groups of different numbers.

■ sorting by given criteria

MM 3 Where will you put it? **IP 3**
1. 'Shut your eyes and think of a very big thing, eg a giant, or elephant, and something very small, eg an ant. Now make a picture in your mind of a big box and a tiny box. Put your big thing in your big box and the tiny thing in the tiny box. Tell us about your pictures. What is in your big box? What is in your tiny box?'
2. Use other objects, especially items from the IP page. Then use other mathematical things such as shapes and numbers. For example, 'Think of two numbers. Put the larger number in the big box. Put the smaller number in the tiny box.' 'Tracy, what is in your boxes?'

■ visualising numbers on a number line

MM 5 Pictures in your head **IP 14**
1. When children have had practical experience using a number line, introduce activities that involve visualising and working mentally with a number line. At first keep this very simple so that children will gain confidence. 'Pretend you are standing on number 1. Now take one step so that you are on 2. Now let's take a big jump. Do it in your head and tell me which number you landed on.'
2. 'Imagine you are standing on 5. Now take one step forward. Open your eyes. Were you standing on this number?' (Hold up a 6 card.) Repeat with steps of 2.

■ visualising a 100 square

MM 6 The hundred square **IP 15**

1. '*Shut your eyes and pretend you are so tiny you can stand on the 100 square on number 3. Tell me what numbers you can see next to you.*' (behind and in front)
2. '*Keep your eyes shut and let's walk right along the top line of the 100 square, 5, 6, 7, 8, 9, 10. I hope you are at the end of the line. What do we need to do to get to number 11?*'

3. '*Open your eyes now and look at the second row. Let's count along from 11. Pretend you have a camera in your mind and you can take a photo of the second row. Now close you eyes and try to see the picture of the second row in your mind.*'
4. Over time, repeat for other rows. By the end of reception aim to reach 100.

Calculating

■ visualising and early language of addition

MM 7 Here come the birds

1. '*Say the rhyme "Two little dickey birds". Now shut your eyes and make a picture in your mind of the two birds. Let's count them, 1, 2. Here comes another bird. You can make your new bird any colour and size you like. Look at it carefully because I am going to ask you to tell us what this new bird looks like. It is going to join the other two birds. Let's count all the birds, 1, 2, 3. Open your eyes and tell us what you saw.*'
2. Repeat with other animals.

■ addition

MM 8 How many now?

● Put out number cards 0 to 5.
1. '*Hold up 3 fingers. Let's count them together.*' '*Which of these numbers is three?*'
   '*If we hold up one more finger, how many then?*' (4)
   '*Can you find the number 4?*'
   '*We started with 3 fingers, added one more and we ended up with 4 fingers. What is 3 fingers add one more finger?*'
   Repeat this with other numbers, adding on just 1 each time.
2. Start with a different number of fingers and do some more adding, using a variety of language and numbers.
   '*Let's start with 2 fingers and add on 2 more. How many altogether?*'
   '*Let's start with 1 finger and add on 2 more.*'
   '*Three and two more makes how many altogether?*'
   '*Two plus two?*'
   '*4 is the answer. What could the question be?*'

MM 9 Introducing addition notation

You need: cards with numbers 0–5, a card with addition sign

1. Use a wide range of words for addition. Write them down and display them. Introduce the plus sign when children have done MM 7.
2. '*Look at this card.*' (The addition sign.) '*Who knows what this cross means when we are doing maths?*'
3. '*Who can read this number sentence with me? Three add/plus/and one more makes 4 altogether*' etc.

■ subtraction

MM 10 How many rabbits? **IP 8**

1. '*Close your eyes and try to see in your mind the 3 rabbits / 4 butterflies / 5 fish. Put them in a straight line in your mind and count them. Now make one run away. How many are left?*'

■ subtraction and comparing numbers

MM 11 Close your eyes **IP 11**

1. Ask children to look at just one of the groups on the page, eg the 4 butterflies. '*Now close your eyes and make a picture in your mind of the 4 butterflies. Let's count them, 1, 2, 3, 4. You can point to them if you want. Now see if you can sit each butterfly on a flower. One flower for each butterfly.*'
2. '*Now make one of your butterflies fly right away. Look at your butterflies still on their flowers. Let's count them, 1, 2, 3. There were 4, but one flew away so now there are only 3. Look at your butterflies and flowers and then open your eyes. At the end, were there more butterflies or flowers?*'

■ number bonds to 5 and 10

MM 12 Five and ten whispers
1. You whisper a number, say 4, to the whole class and everyone must hold up the number of fingers that makes 4 up to 5 (1). If children play this in pairs, one whispers the first number and the other whispers back the number to make it up to 5/10.
2. Before children leave reception develop this up to 10 so children begin to memorise number bonds to 10. For example 7 and 3. Follow this with reversing the numbers, 3 and 7.
3. You can repeat this with numbers other than 5 and 10.

MM 13 Get on the bus **IP 6**
1. You will need to do this activity several times, developing the language each time and repeating the number bonds with CG 32 and MM 12.
  'Shut your eyes. Imagine 5 children. One goes upstairs. How many go downstairs?'
2. *'In your head, picture 10 children. You can put as many as you want upstairs in the bus. Tell us what you did.'*

■ strategies for addition and subtraction

MM 14 How did you do it? **IP 10**
1. When you have done a lesson with IP 10 you can keep practising at mental maths time, eg calculating how many children are away, how many are having packed lunches, etc. Always ask children *'How did you do it?'* At this stage you are getting children used to talking about what they do mentally and the pictures in their head, so you need to accept almost every method. These are likely to vary enormously and may not involve mathematics, eg 'I knew it was 6 for school dinners because Kevin is away,' and 'You wrote 13 on the dinner register so I knew it was 13.'

Shape

■ imagining shapes

MM 15 Picturing shapes **IP 9**
1. Read and do the actions for 'Imaginary pictures' *Seven dizzy dragons*, page 9.
2. Show shapes on IP 9 and later use plastic or wooden shapes (2D and 3D). Ask children to look at one then shut their eyes and make a picture of it in their mind.
3. *'Look at this shape (cuboid shape). Now make a picture in your mind of lots of these shapes and build something with it in your picture. Tell us about what you built.'*
4. *'This is another shape (empty cuboid box with lid) that has 6 flat sides just like the brick but has a space inside it. Make a picture in your mind of a shape like this.*

You can make it as big as you want. It can be enormous. Now put something inside it and shut the lid, if you can. Maybe the thing you put inside it is so big you can't shut the lid. Tell us what you put inside.'
5. *'Here is another shape with 6 flat faces (cube) but look, this time, all the 6 faces are the same size. Now think of a centimetre cube like this one. Keep that tiny cube in your mind, then think of a cube that is a bit bigger and has a lid that you can open. Open it up and put the tiny cube inside. Now make a slightly larger cube again with a lid that opens. What do you think I am going to ask you to do? Yes, put the other cubes inside this cube.'*
6. Repeat with other shapes, 2D and 3D, and other sizes.

Circle games

These games can be played with the whole class sitting in a circle on the carpet, often with children taking turns, going around the circle. They are designed to last for just a few minutes, to support your daily whole-class mental maths work right through nursery and reception.

Most games involve some apparatus so that children are working with objects in the early stages of their mental maths work. This is listed for each game. A clear number track is needed on the floor and wall for many activities and large hundred square (IP15) is also helpful.

Although many of the games are for supporting children in the nursery, working towards level 1, most can be extended through levels 1 and 2 and some into level 3.

Sorting and matching

■ sorting into groups

CG 1 Sort it out **N(ursery)**

1. Give children 2 or 3 options of groups, eg use 2 large bits of paper, one labelled 'numbers' and the other 'letters', to see if children can sort numbers from letters. Hold up one number or letter at a time and ask children which group to put it in, or give out numbers and letters and go around the circle, sorting each one into the appropriate group.

2. Use other mathematical equipment, eg sort out triangles and squares, or 'bricks good for building' and 'bricks that roll', 'numbers less than 5' and 'numbers 5 and more'.

Follow on

3. When children are more experienced, let them decide on their groupings for the objects.

> **You need:**
>
> objects, eg three small blue sorting toys (a car, house, elephant), a red set of similar objects; some bricks, Lego, Polydron etc; some triangles and squares in red, blue and yellow. In later nursery use more mathematical things such as wooden numbers, dotty cards, 2D or 3D shapes.

■ sorting and matching

CG 2 Go togethers **N**

Children sort and match as a part of their play and they are usually good at it by their second year. This circle game helps to develop their skill in mathematical contexts.

1. Children try to pick up an object that goes together with the previous object, eg 2 Polydrons of the same shape or colour.

2. Ask children in turn to pick up a suitable object from the centre of the circle and describe it, eg 'I've got a yellow elephant.' The child can be encouraged to talk about their choice and why it is a 'go together', eg another elephant, or the same shape or colour.

3. This very basic game can be developed in many ways, eg matching 3D shapes, numbers, see CG 14 Find a partner.

> **You need:**
>
> sorting toys as in CG 1.

Numbers and the number system

- **recite the number names in order from zero**

CG 3 Finger counting **N/R(eception)**

1. Say a number rhyme. Count around the circle, with children in turn saying the next number. If you want to start counting from one again, explain that you are starting the numbers again. '*You start from one again Sammi.*' Even if children cannot count out objects reliably they can learn to chant the numbers in order. Move on to ten once children are aged approaching four years old. Start with zero once children have some basic knowledge of numbers.

> **You need:**
> number cards 0–5/10/20

CG 4 Pass the parcel **N/R**

1. Pass around a parcel saying the numbers as the parcel moves from child to child. Emphasise that you say one number word for each child.

Follow on R

2. This time pass the parcel around missing every other child, starting on zero. Ask children to predict if number 10 will get the parcel? Start the parcel, saying the numbers, 2, 4, 6, 8 etc. Repeat, this time asking about another number, eg '*Will number 5/8/20 get the parcel?*' Then start at number one, not zero, and again ask for predictions.

> **You need:**
> number cards or sticky notes 0–30 (all the class) stuck onto jumpers

- **recite the number names in order, counting on from a given number**

CG 5 Beads **N/R**

1. '*I'm holding up 1 finger. Now one more. How many is that? Do you need to count them?*' Do plenty of finger counting and number rhymes and then try to establish that children do not need to count from one each time.

2. Focus on the five fingers of one hand and give repeated experiences with one hand with all five fingers and some of the fingers in the other hand. Many children will still want to count their 5 fingers again each time, so ask: '*Do you really need to count those fingers again? How many was it? Yes, five. Now try to keep that hand up and two fingers on the other hand. We don't need to go back to one. We know this is 5, so: "six, seven".*'

3. Count with beads for a change, using 5 of one colour, and 5 of another.

> **You need:**
> lots of beads, IP 8

Follow on R

4. Count around the circle to 5/10 and then ask children who will be number 5/10 when you start again. Give children time to work it out.

5. Put children in pairs so that they have 20 fingers between them. *'Show me 13.'* Demonstrate with one child holding up ten fingers and the other 3. *'One ten and 3 more fingers makes 13 altogether.'*

6. Remind children that they don't need to count the ten fingers each time.

7. Emphasise the ten and some more when you are counting, so 14 is ten and four more, 15 is ten and five more.

8. Complete the numbers to 20 on the number track on IP 8.

CG 6 Count on, count back **R**

1. Generate numbers in a variety of ways: pick a card, spin a spinner, or throw a dice.

2. Shuffle the 'count on' and 'count back' cards. A child chooses one to see if they count on or back from their number.

3. If a child has 'count back' and spins 7, they must say 'I start at seven, then six, five …' etc.

4. Give the number with irregular beats on the drum or claps. At first let the child count them with everyone else.

5. Repeat the game as above, but just counting back, generating numbers above, say, 5. Have a second set of cards, say 0–4, giving the number to stop at. For example, *'Count back starting at 7, stopping at 4.'*

6. Extension
'Can you count back from your house number all the way to zero?'
'What is the largest number you know? Can you find it on the number line and show us at review time how you counted back from that number?'

> **You need:**
> two cards labelled 'count on' and 'count back', number cards, spinner or dice, number line on display

■ **knowing number names for larger numbers**

CG 7 Collecting a hundred **R**

1. Over a week collect a hundred of something. At the end of each day count how many you have, circling the number on a hundred square or a 1–100 number line. *'We have 46 buttons now. Let's find that number on the hundred square.'*

2. When you have 100 items, group them into 10 lots of 10, and label each container.

> **You need:**
> a hundred of something, eg paper clips, Unifix cubes, buttons, IP 10 or IP 15

Follow on

3. Relate some of the larger numbers that children are aware of (eg house numbers) to a number of items. Children who can read 67 for example, as their house number, will be thinking of 67 more as a label than a quantity, so counting out 67 items will give experience with quantities.

■ **count reliably a collection of objects (one-to-one correspondence)**

CG 8 Move and count **IP 1 N**

1. Give each child 3 pebbles. *'Count them carefully. As you say each number name move one pebble.'* *'Now put your pebbles in a different pattern* (eg all in a long line). *How many now?'* Continue making different patterns to emphasise that the number is the same whatever the arrangement.

2. *'Let's count the bears on IP 1: 1, 2, 3. Count out that many pebbles / cubes / teddies. Let's move them as we count them: 1, 2, 3. How many? Hold up that many fingers.'*

3. *'Who will show us how they count their teds, moving them and counting out loud?'*

Follow on R

4. Use more pebbles and different pictures on IP 1 to count up to other numbers, eg 4 rabbits. Match the numbers to fingers, to the number line on the wall, to number cards, dotty cards with dots in different arrangements, and give experience with counting up and down as steps along a floor number track.

> **You need:**
> pebbles, cubes, teddies

CG 9 Harder counting **IP 13 R**

1. Draw items randomly, eg 7 dots on the board on IP 13. Ask children to count them although they are out of reach. Ask a confident child to check by counting them by touching.
'How will we remember where we started?' 'Have we already counted that one?' 'What could we do to make sure we don't count something twice?'

2. In pairs or small groups, children take a handful of Lego / teddies. *'Count your bricks and agree on the number.'* Watch for strategies such as moving and counting, or counting in twos etc.

> **You need:**
> fairly large objects for counting, eg Lego, teddies

■ **count reliably in more difficult contexts**

CG 10 Count and clap **N**

1. *'Everybody say: "1, 2", clap, clap, "1, 2", clap, clap.'* Gradually recite more of the numbers, eg '1, 2, 3, 4 (then 4 claps) 1, 2, 3, 4 (4 claps)'.

Follow on R

2. Without clapping count up to and beyond 100. (See also CG 38 Copy me.)

3. Ask a child to skip while the rest of the group count how many times they jump.

4. Use a chime bar or heavy balls that will drop noisily into a metal container, out of sight. Make a number of sounds to a regular beat, up to about 20, and ask children to count silently in their heads. Make irregular sounds for lower numbers.

■ conservation of number

CG 11 Bricks and fingers **N/R**

1. *'Count out 5 bricks each. Put them in front of you. Put one brick on the tip of each finger. Do you have one brick for each finger?'*

2. *'Count the bricks, moving them as you count. Put the bricks behind your back so you can't see them. How many bricks do you have behind your back?'*

3. *'Put your bricks in a long line in front of you. Count them. How many now?'*

4. *'Now make a tower. How many?'*

5. Emphasise that whatever we do with them, there are 5 because we aren't getting any more out of the box, and we aren't putting any into the box.

Follow on R

6. Arrange children in groups of 3, 4 or 5. They say how many are in their group. The children in each group hold hands and dance while the music plays. Stop the music and ask each group *'How many in your group now?'* *'No-one has left your group. No-one has joined your group. There are still 5 children. Let's count your group very carefully.'* Help children to count around the circle. *'Make sure you don't count someone twice. We started at Kate so we must end at Jay.'*

> **You need:**
> Unifix cubes that can make a 'hat' on the end of a finger, or small bricks

■ count in tens/twos

CG 12 Don't count in ones! **IP 1 Late R**

1. Counting in 2s is quite sophisticated and children will do it by rote for some time. Count the animals going into the ark on IP 1.

2. Stand a couple of children in the middle with their hand held out. *'How many hands? Don't count in 1s.'* Guide children to counting in 2s, pointing to the hands, *'Two, four.'*

3. The next time you do the activity, ask children

> **You need:**
> Multilink, IP 10, IP 15

to say how they did their count and to recite the counting in 2s again as a whole group. As children get more experienced in this, use up to 10 children or draw some pairs of dots on the board before the session. Counting in 2s just with one different set of dots for each day (eg on Monday display 5 rows of 2) could consolidate this work and the dots can be left on display during the day.

4. When children are confident, go around the circle counting in 2s as far as children can go. Use the hundred square on IP 15 or the number line on IP 10 to help to see the pattern and make the link to hopping in 2s.

Follow on

5. You could extend this game to counting in tens using children's ten fingers or Multilink made into 'ten trains'. You may find some children count in 5s; if so develop that. Circle the sequence of numbers on the hundred square on IP 15.

■ Begin to read numbers and number names

CG 13 Take a number from the pot **N**

1. Ask a child to take out one card. '*Hold onto the felt strip and feel the number with your finger.*' Help if necessary. '*What number have you got?*'

2. '*Count out the appropriate number of cubes or teddies.*' Other children can help.

3. Call children up in turn, saying this rhyme together as you do so:

> Take a number from the pot
> then sit down,
> what have you got?

> **You need:**
>
> Tactile number cards zero to 5/8/10/15/20, a pot, cubes. Have a clear number track to 10 or 20 on display, eg IP 2 or 8

CG 14 Find a partner **N**

1. Lay the number cards out in sequence. Hold up a dotty card. '*How many dots?*' '*Come and find the number card that goes with this.*'

2. When children are confident, play it around the circle. Give out the dotty cards, and as you go around the circle children must match their dotty card to a number card in the middle of the circle.

> **You need:**
>
> number cards, dotty cards in standard patterns and cards with written number words

Follow on R

3. Hold up a card and ask children to respond immediately to the number by holding up that many fingers. Count them with the children.

4. Match the numerals to words for numbers instead of dots.

■ begin to understand and use vocabulary of comparing/ordering numbers

CG 15 How many more is your number? **R**

1. Pair up the children round the circle. Put counters within reach of them. Give each child a number card. Children count out their number of counters.

2. *'Now compare your number with your partner's. Who has the biggest number? How many more is your number than Josie's?'* Let children take turns to talk about their number, eg 'I have 3 and it is more than Kelly's number 2.' 'I have 4. It is less than Jake's because he has 7.'

Follow on

3. Order some of the numbers into a number line and talk about these. *'Three comes here before four because 3 is less than 4.' 'Tell me the number that is one more than 5.'*

> **You need:**
> dotty cards, number cards, things to count with

■ order a complete sequence of numbers

CG 16 Washing lines **N/R**

1. *'Let's put these number cards in order from 1 to 5 and hang them on the washing line.'*

2. Recite the numbers in order. Leave that line on display.

3. Call out 5 children. Give each one card from a 1–5 set. *'Put yourselves in order.'*

4. Repeat with other sets of cards and children, doing the activity in small groups.

5. Repeat without the washing line on display.

Follow on R

6. Count along the washing line every day for several weeks. Gradually add more numbers and try to get at least to 15 and then 20 by late nursery. Children will gain from the chant of the counting and the sounds of the number words even if they cannot count out 20 objects or even count independently to 3.

7. Use the number track on IP 8 or the roller coaster on IP 9 to write and order numbers up to 20 and beyond.

> **You need:**
> 1–5/10/15/20 number cards, pegs and a 'washing line' (At first do the activities with a second washing line or number line constantly on display for matching.), IP 8, IP 9

CG 17 Washing in the mud **N/R**

1. With the number cards in order on the washing line, say the numbers in order several times.

2. *'Close your eyes.'* Reverse one or two of the cards so that the number is hidden. *'Open your eyes.'* Point to a reversed number *'Which number is this?'*

3. Lay the number cards in order face down on the
    carpet. '*All the washing has fallen off the line but it is
    still in order.*' Point to one of the cards. '*Which
    number is this?*' '*Can you find me number 3/4/7/10?*'
    Repeat this daily for several weeks.

Follow on

4. Move one of the cards out of order. '*Which
    number is in the wrong place, and where should it
    be?*'

5. Repeat any of these activities without a number
    line on display.

■ say a number lying between two others

CG 18 Caterpillar count **IP 1 N/R**

1. Lay the caterpillar head down and invite
    children to count with you as you or a child put
    down more circles for the body.

2. Say the following rhyme, putting more circles
    down as you do so.

    > *Caterpillar, caterpillar*
    > *how big will you be?*
    > *Find some more circles*
    > *then we will see.*

3. Number the circles in order. '*What is one more
    than/less than 5?*' '*Which number comes just before
    3?*' '*What comes next?*'

Follow on

4. The caterpillar can be put on display and can
    act as a number line. You can add to it over the
    next few days.

> **You need:**
> one circle with a face on it for the caterpillar's
> head and 5/10 circles of paper

■ order some selected numbers

CG 19 Empty number line **IP 14 Late R**

1. Choose a child to be zero. Stand another child a
    little way away from zero and call that child
    number 10. Put a pot numbered 10 at her feet.

2. '*Who would like to put this pot 5 at about the right
    place between zero and 10?*'

3. Establish that this pot is about half way. Put in
    other numbers.

4. Repeat with other numbers, eg zero and 5/15/20.

Follow on

5. Use the empty number line at the bottom of IP
    14. '*If the ted with the ball is near number 10 and the
    ted hiding in the bush is zero, where should we make a
    mark for 5?*' Continue with other numbers.

6. '*If the ted with the ball is at a hundred, can we mark
    in some other numbers?*'

> **You need:**
> clearly numbered yogurt pots

■ **estimate by choosing the more likely of two numbers**

CG 20 Don't count, just guess! **R**

1. Hold up a card. '*Are there about 5 dots or about 20 dots? You don't have to count them!*'

Follow on

2. Use other things around the classroom, eg a large box of Unifix. '*Is the number of cubes in the box nearer to 5 or nearer to 100?*' '*Are the number of people in our class nearer to 30 or to a million?*' Over the top alternatives make it easier for children to appreciate which is more likely.

3. Use IP 3: '*Are there nearer to 5 clowns on this page (4) or nearer to 20?*' '*Are there about 10 stars or about 30?*' (There are 14 stars.)

> **You need:**
>
> cards with dots arranged randomly, a few with about 20 dots (eg 19 and 21) and a few with about 5 (eg 4 and 6)

■ **begin to recognise and use halves in context**

CG 21 Half ran away **R**

1. Make the link to halving numbers by reading 'Twelve little monkeys' *Seven Dizzy Dragons* page 27. Counting out cubes and putting them into two groups to find half will be an essential practical activity. '*Double 5 is 10, look at my fingers, now I can find half of 10, it's 5.*'

> **You need:**
>
> cubes

Calculating

■ **begin to understand and use the vocabulary of addition and subtraction**

CG 22 Adding with cards and teddies **R**

1. Seat the children in pairs round the circle. Give each child a number card. They take that number of teddies from the centre of the circle.

2. '*Turn to your partner. Keep your two sets of teds separate but find how many you have got altogether.*' Go around the circle and ask each pair to show their two number cards. Help them to say a number sentence based on their numbers, eg '*Harry has 4 teds and Abdul has 2. 4 teds and 2 teds makes 6 teds altogether.*'

Follow on

3. Introduce cards with + and =.

4. Progress to writing sums using the 5 boxes at the bottom of IP 10.

> **You need:**
>
> several sets of number cards 1–5, lots of teddies/ pebbles, IP 10

■ addition as the combination of 2 groups of objects

CG 23 Lily pads **R**

1. Put 3 lily pads in front of you. Sit 3 frogs on the first lily and one more on the next. *'All the frogs want to be together so they hop onto the third lily pad.'* Move all the frogs. *'How many are there altogether?'* Use a varied vocabulary of addition.

> *'Three here, then one more makes four altogether.'*
> *'Three and one more makes four.'*
> *'If we add 3 and 1 we get 4.'*

2. Group children and give each group some frogs. Then go around the circle helping children to make up a mathematical sentence about their frogs, eg *'I had 2 frogs and then I got 4 more. That makes 6 frogs.'*

Follow on

3. Try with four lily pads for addition of three numbers.

> **You need:**
> frogs (cubes) and lily pads (paper circles)

CG 24 Train ride **R**

1. Going around the circle, pass the train with nothing in any of the carriages to the first child. They put a few cubes into the first carriage and tell everyone how many are now on the train. That child drives the train to the next child who puts some into the second carriage and says how many passengers they have added. The third child combines the passengers and says how many there are altogether. *'3 and 2 more makes 5 altogether.'*

2. The passengers are now counted off at a station, and the activity can be repeated.

Follow on

3. *'Which number comes after 3? Get one more ted. How many now? Hide one behind your back. How many in front of you?' 'Turn to the person next to you and put all your teds together. Four teds and four more. Let's count them together.' 'Put all 8 teds in a row. Pass the box around the circle and put 2 (3/4) of your 8 teds in the box. How many teds will you have left?'*

> **You need:**
> teddies (cubes) and a 'train' with an engine and 3 carriages (eg margarine pots joined with string)

CG 25 Dominoes **R**

1. Pass around the pot. Children take one domino, say the two numbers, add them, give the total and pass the pot on. Try to use the varied language of addition.

> **You need:**
> a pot with all the dominoes that add up to 1, 2, 3 and 4 (Take out the rest, but repeat the game several times over the year, gradually including larger numbers.)

Follow on

2. Take out all but the dominoes that add up to 5, 6 and 7. Call this game 'Making 5, 6 and 7' so that children know what to expect.

3. RS 3 can be used for recording.

■ **addition as counting on**

CG 26 In your head **IP 2 R**

1. Developing the mental image of a number line is important for calculating, eg '*Here is number 5 (point to it on the snake on the IP). Add 1 (take a step with your fingers). What number do I land on?*'

2. Repeat that kind of question but move gradually from the language of physically taking steps, eg '*Which number do I land on?*' to the more abstract language of addition, eg '*What is 4 and 2 more?*' '*6 add 1 more. How many?*'

■ **addition of doubles by counting on**

CG 27 I can double numbers **R**

1. '*This is double 3.*' Hold up 3 fingers on each hand. Emphasise that doubling means two of the same number.

2. Work through all the numbers to 5, using fingers, cubes and other counting objects.

3. '*Show me double 1/2/3/4/5 with fingers. How many altogether?*'
Repeat this frequently in mental maths time, moving on to double 6 when appropriate.

4. Progress from objects to using dotty number cards or a dotty dice. Children, in turn round the circle, take a number/toss the dice and tell everyone the double. (Some may find it helpful to continue using fingers.)

Follow on

5. Use domino doubles at mental maths time. '*Here is double 6. How many dots altogether? Shut your eyes and imagine the dots on the domino and count them in your head.*'

6. Put double dominoes in a bag and pass it around the circle. Children pull out a domino and say what their number is, eg '*I've got double 2 and that is 4.*'

You need:	
cubes	

- **separate a number of objects into two groups**
- **combine two groups to make a particular number of objects**

CG 28 Inside outside **IP 11 N/R**

1. Give each child 3 stones. *'Let's count them together, pointing to each one in turn.'* Ask children to hold up 1 finger for each stone and count them.
2. Put one stone inside the circle so there is one inside and two outside. *'Do this with one of your stones.'* Children can then copy what you are doing and talk about it. *'How many are inside the circle?' 'How many are outside?'*
   'Are there more inside, or more outside?'
3. Change the arrangement. Include 3 inside and none outside, as introducing the concept of zero early is important.
4. Ask individual children to make a new arrangement. *'How many are inside? How many outside?' 'If you take one more out of the circle, how many will you have outside then?'*

> **You need:**
> pebbles/shells/cubes/teddies, quoits/hoops/boxes

CG 29 Splitting numbers **IP 11 R**

1. Sing 'Five little speckled frogs'.
2. *'Five frogs altogether.'* Ask 2 children to stand in the hoop, 3 just outside it. *'2 in the pond, 3 out. How else can we arrange the frogs? For example, 1 in and 4 out.'*

> **You need:**
> a pond (a large hoop) and frogs (cubes)

Follow on

3. Children in pairs could play this game with toys and record their combinations on RS 5.
4. Repeat this activity in different contexts, eg using the fingers on one hand, some up, some down, children inside and outside the playhouse, jam tarts, some in the oven, some outside already cooked.
5. Play grouping currants on two sides of a bun. Put RS 6 on the maths table for children to record as a teacher-independent activity.
6. Progress to using larger numbers.

CG 30 Dragon is hungry **IP 16 R**

1. Invite one child to be the dragon who is very hungry and another to be the clown who doesn't eat very much. *'They have gone out for a picnic and have taken 5 buns to eat. How can they give them out so that dragon has more than clown?'* Talk about the various arrangements of the buns, putting them on two paper plates. *'Are there any more ways we could give them out so dragon has more?'* (5 for dragon, 0 for clown; 4 and 1; 3 and 2.)

> **You need:**
> 2 paper plates, buns (cubes)

2. With children working in pairs, repeat the activity with 5 buns. *'Dragon must have more buns than clown each time. You will need to count them to check dragon has more.'*

Follow on

3. Children can investigate ways of splitting other numbers and splitting numbers between 3 or 4.

CG 31 Teddy's front door game **IP 12 R**

1. Sit the children in pairs round the circle. Hold 2 teds in front of you and hold 4 behind your back. *'I have 6 teds in my hands. You can see these 2. How many are hiding behind my back?'*

2. Give each child 6 cubes. *'Split your 6 into 2 groups.'* Ask a few children to say what they have done. Repeat the answers to bring out the mathematical language. *'Louise had 6 cubes and split them into 3 and 3.'* Establish that 6 can be split into two groups in several ways.

3. Look at the IP. *'Ted likes to play a game with front door numbers. He chooses a front door number and puts it on the door, eg 6. He then splits 6 into 2 numbers for the windows. What could those two numbers be?'* Write the numbers on the windows.

4. Give each pair a ted house and ask them to count out 7 cubes. Write 7 on the front door on the IP. *'Split your 7 cubes between the two windows.'* Ask children to explain what they did.

5. *'How many will you have if you put your two groups back together again?'* Hopefully some children won't need to count them to know there are seven! *'Amy, how did you know there were 7 so quickly?'* Emphasise: *'So you knew you didn't need to count them.'*

Follow on

6. *'With 5 on the door and 2 in this window, what is the number for the other window?'*

7. Split the number on the door into 3. Record on RS 8.

■ **number bonds to 5/10**

CG 32 Make up to 10 **R**

1. Shuffle the cards and give one to each child. *'Work out what number you need to make your number up to 5.'* Give children time to think. Everyone holds up their card. *'Look to see who has the number you need. Who would like to invite that person to come and sit next to them?'* *'Darren has a 3 and I have 2 so please will Darren come and sit next to me?'*

> **You need:**
> teds (cubes), a paper house with windows and door as on IP 12 / RS 8 for each pair, and number cards to fit on the door and windows

> **You need:**
> number cards 0–5 (later 10)

Follow on Late R

2. Use different target numbers. With even target numbers you will need to include two of the half-target card, eg for 0–10 you need two 5 cards.

3. Try with numbers to make 20, selecting just a few pairs at first eg, 10 and 10, 5 and 15, 19 and 1.

■ remove a smaller number from a larger and count the remainder

CG 33 Start with … **N/R**

1. Each child takes a handful of bricks. *'Count your bricks.'*

2. *'Take turns to throw the dice. Take that number of bricks away. How many have you got left?'*

> **You need:**
> bricks, a dice or spinner (0, 1, 2, 3, 3, 3)

CG 34 Crossing out **IP 8 R**

1. Count items with children on IP 8. *'There are 5 fish but this one swims away. We can cross it out like this. How many fish are there now?'*

2. *'There are 6 cats. Everyone hold up 6 fingers. One ran after a mouse. Put one finger down. How many cats are left?'*

Follow on

3. Have a daily session on crossing out for several days in mental maths time, using IP 8 or IP 11, and link this with a wide range of language of subtraction.

■ find out how many have been removed from a larger group of objects

CG 35 Mice in a hole **IP 13 N/R**

1. *'I've got 5 mice. I'm putting one of my mice behind my back. How many are left in front of me?'* (4) *'All my mice hide behind me. These 2 creep out. How many are still hiding behind me?'* (3) *'How did you do that? How do you know you are right?'*

2. Children play the game with a partner. Ask them how they work out their numbers.

> **You need:**
> lots of mice (cubes) and a cloth for a hole

Follow on R

3. *'Shut your eyes and make a picture of your 5 mice. One of them hides. How many can you see now?'*
'How many mice would I have if I had 2 in front and 2 hiding?'
'How many is two mice and one more mouse?'

4. Give plenty of mental maths practice. *'There are 3 mice in the hole. Another joins them. How many now?'* *'There were 5 mice in the hole. Now there are only 3. How many ran away?'* *'There are 5 mice in the hole. How many need to go in to make 6 in the hole?'*

5. This game is repeated with other numbers of mice through the practice books and on RS 4. Aim to work with 10 by the time children are in their last term of reception.

6. Work towards systematic recording by asking children if they can do the recording in order.

■ work out how many are needed to make a larger number

CG 36 How many more do we need? **R**

1. *'Anna has 3 acorns. How many more does she need to make her number of acorns up to 5?'*
'Let's start counting at 4 and go on until we get to 8.'

2. *'Shut your eyes and think of 4 robots at a party. Some more come to join them so that there are 6. How many more came to the party?'*

Follow on Late R

3. By the end of reception, emphasise the number bonds to 10.
'Jos has 6 shells. How many more does he need to make 10?'
'Sasha had 10 pennies, but she spent some. She only has 2p left. How much did she spend?'

4. *'Hold up your hands. How many fingers have you got? Put down 3 fingers. How many are up? If we put 7 down, how many will be up?'* Try to establish this reversing pattern, 6 up, 4 down; 4 up, 6 down. Link these to pairs of number cards.

5. Show 3 fingers. *'Hold up the number of fingers that will make this up to 10.'* Play this in pairs.

> **You need:**
> things to count

■ find the difference between two numbers by counting on or back

CG 37 Take away **N/R**

1. Sit the children in pairs round the circle. *'There are 3 sheep in the field but 2 got out. How many are left in the field?'* Repeat this with other numbers.

2. Give each child a number card. *'Put that many animals in your field. Compare the number of animals you have with your partner's number, and decide who has more and who has fewer.'* Ask pairs to tell you what they found. *'I've got 4 teddies but Jess only has 2 so I have more.'*

3. *'Tim has more than Jess. How many more? How can we work this out?'* (Matching animals and counting the number that can't be matched.)

4. Circle Tim's and Jess's numbers on a number track. Draw in the steps from the smaller to the larger number. *'This is the same as the difference we found by counting. Would someone else like to try their numbers?'*

> **You need:**
> a field (paper, quoit) and small sorting toys for each child, number cards

Follow on R

5. Children can compare their ages with those of siblings, eg *'Your little sister is 3 and you are 5. What is the difference in your ages?' 'The difference between 3 and 5 is 2.'*

■ copy, make, describe and continue simple patterns

CG 38 Copy me **N/R**

1. Clap a pattern for children to copy, eg quick, quick, slow, quick, quick, slow.

2. Go round the circle with children doing consecutive elements of a pattern, eg tap knee, stamp a foot, tap knee, stamp a foot.

Follow on

3. Play Follow my leader in PE making patterns, eg step, jump, step, jump.

Making sense of number problems

■ solve simple number puzzles in a practical context

CG 39 Share out the apples **R**

1. Sit the children in pairs. Ask each pair to count out 4 apples and share them fairly. *'How many each?' 'Now try with 6 apples. How many each? Did you have any left over?'*

2. *'Try it with other numbers (to 10)'. 'Which numbers can be shared so you each get the same number of apples?'*

> **You need:**
> apples (cubes)

■ understand and use vocabulary related to length/mass/capacity

CG 40 Red Riding Hood's basket **IP 4 N/R**

1. Put all the objects in the middle of the circle.

2. Pass around the basket. Ask children to put items in: *'Put in a short ruler.' 'Put in a large ball.' 'Put in something heavy.' 'Put in something large but light.'*

3. When the basket is full, ask children to take things out in the same way.

> **You need:**
> long/short, large/small, light/heavy objects

CG 41 Ribbons **IP 3 N/R**

1. *'Mr Giant is going to buy some ribbons. He wants the longest bit of ribbon he can find.'* Lay out the ribbons side by side in the middle of the circle. *'How can we find out which one is the longest?'* Let children compare them. Introduce words such as *'longer than, shorter than, longest, order'. 'So*

> **You need:**
> ribbons of various lengths easily distinguished by colour

which one do you think Mr Giant should buy?' 'If he wanted two long ribbons, which is the next longest?'

Follow on

2. *'Baby bear wants the shortest bit of ribbon and mummy bear wants a medium sized bit. Find ribbons for them.'*

CG 42 Get in order **IP 6 N/R**

1. Ask three children to stand in the middle. *'Work out who is tallest. Who is shortest? Who is the middle?'* Ask them to make a line, in order of height.

2. If three children can be ordered without much trouble, ask five to stand in the middle.

Follow on

3. *'Pretend all the dolls/soft toys are at school and they have to get into a line starting with the tallest/shortest like you did.'*

4. Talk about height in other contexts, eg *'Do you think the cupboard is taller than the piano? How could you find out?'*

Shape and space

■ properties of shapes

CG 43 Match my shape **N/R**

1. Give out plastic 2D shapes. *'Look at your shape. Count how many sides it has. Are any curved? ... Hold up your shape. Look to see if anyone else has the same shape as yours. It doesn't have to be the same colour and size.'*

2. Ask children in turn to describe their shape and invite someone with the same shape to came and sit by them. *'My shape has 3 straight sides like Nadia's. Please come and sit next to me, Nadia.'*

Follow on

3. *'Hold your shapes in your laps secretly and Lisa, please describe your shape without showing it. If anyone else thinks they have the same shape as Lisa, put up your hand. Compare your shapes to see if you were right.'*

■ halving and vocabulary of symmetry

CG 44 Fold it up **R**

1. Give each child a paper shape and ask them to fold the paper in half. '*Hold up your shape. Look to see if anyone else has the same shape as you.*' Ask children in turn to describe their half-shape and invite someone with the same shape to join them.
2. '*Fit your two half-shapes together so they make the shape you started with.*'

Follow on

3. Make 'blob' paintings, fold in half and press to make symmetrical butterflies or monster faces.
4. Give children mirrors to play with on the maths table and a variety of shapes and pictures, some with a missing half.

> **You need:**
> paper squares, circles, equilateral triangles

■ language of position

CG 45 Where is it? **R**

1. Direct children to put their toy into positions using a variety of words such as 'on top of, under, in front of'. '*Put your toy beside your knee/on top of your head.*' Start with simple directions.
2. '*I'm putting my ted between my feet. You put your ted between you and the person next to you.*'
3. Once children are confident with the words you can go around the circle asking children to make a sentence about their toy, eg '*My tractor is going along my arm.*'

Follow on

4. Ask individual children to choose a word, eg 'behind', and demonstrate it with their toy.

> **You need:**
> a small toy and box for each child

■ recognition of solid and flat shapes

CG 46 What's in the bag? **R**

1. Choose a shape and talk about it. '*Is it flat or solid?*' '*Let's count the sides.*' '*This shape has a curved edge.*' '*This is a small shape with 3 straight sides.*' '*This brick has a curved face but this one has flat faces.*'
2. Put a few items in a feely bag, and pass it around the circle. '*Hold one shape. Can you describe what it feels like? When someone thinks they can identify it they say its name and you pull it out.*'

> **You need:**
> bags, boxes or pots, a selection of coloured 2D or 3D shapes of different sizes and colours, eg Logi-blocks

Structuring play for numeracy

This section shows you how to set up structured play activities to have a numeracy focus.

Play is an important activity in the early years classroom as it is the means by which young children learn. Whenever children are playing, they are learning and developing language. Consequently, as well as setting up structured play activities we should also value, observe, assess and build on the 'free flow play' that surrounds us each day and constantly amazes us.

Structuring a play session

Play fits well into the three-fold plan-do-review (starter, group time, review) lesson structure for those children you want to work fairly independently of you at a particular time.

1. Plan

Children can either
- make an entirely free choice of their play
- or be given an area of the classroom to work in and they decide at the start of the lesson what it is that they want to do there
- or be given a specific task to do by you, usually, but not necessarily, something to do in a specific area of the room.

Children plan what they are going to do either on paper through drawings and emergent writing or simply by telling the rest of the class what it is that they are going to do.

Questioning can help children in their planning process. For example:

'What do you think might happen?'
'Why do you think that?'
'How are you going to find that out?'
'We will ask you about ... at review time.'

2. Do

During the group time let children play fairly freely. Scan them, listening and observing when you can, so that you can ask good questions at review time. Although visiting a group of children who are playing can disrupt them, your sensitive intervention can sometimes be a great catalyst for developing thought and play, moving children towards the outcome you are aiming for. You could ask:

'What are you trying to find out?'
'Tell me what you are doing?'
'Could you do that a different way?'
'Please bring that to review time and tell everyone about it.'

3. Review

Ask children to say what they planned to do as well as describing what they actually did.

Your questions need to:
- encourage language development
  'So tell us why you think that happened.'
  'Did you expect that?'
- tease out important concepts that have been learnt
  'So Amie found out that ... What do you think about that Jo?'
  'I remember Teri did something like that last week when she found out that...'
  'I wonder if that fits in with what we are learning about at the moment?'
- highlight areas for further investigation and learning that you and the children will address in the next few days or weeks. (Often, you can follow these up by making an interactive display so that children can go on investigating and you can continue talking about it in class.)
  'Maybe tomorrow someone working on the big blocks will be able to tell us more about that.'
  'What do you think we need to write up/remember for tomorrow?'

There are suggestions for specific questions to ask for some of the maths table and other tasks in the lesson plans.

Activities to raise awareness of numbers in the nursery

- Use notices and labels around the room and read them regularly to the children, eg *'2 children can play with the doll's house,'* *'Today is Tuesday so everyone who is 4 goes to big school assembly at 11 o'clock.'*
- Keep charts and lists around the room, eg *'4 teachers today – 3 coffees and 1 tea at break time,'* *'34 children here today,'* *'Tick your name when you have made jam tarts.'*

- Make interactive displays of number rhymes which have items to stick on with Blu-Tack, together with related number cards.
- Have a large floor number track and count along it several times a week.

The activities below are set in contexts familiar to children and may be set up in the classroom. You will find ideas for short games and 5-minute activities in the mental maths and circle games section.

The main mathematical learning objectives are given for each activity, but these are obviously not the only learning outcomes that could be achieved through doing this activity. Some of the many other skills and concepts that could be developed by altering the focus of the activity are sometimes given. The activities have been grouped according to the most prominent learning objective.

Measures

■ language of comparison of measures

P 1 In the shop

- *'You are stacking the shelves. Arrange the things in order on the shelf. Later you can tell us about the order you put them in.'*
  One order could be from the tiniest toy in the nursery to the largest one, or the big green bucket that holds the most water to the doll's bottle that holds the least.

Other contexts: at the clinic, the tallest doll to the shortest.

> **Provide:**
> plastic bottles of varying heights or capacities, heavy and light objects, containers in the sand and water, tall and short dolls, soft toys with different kinds of fur, colours and hats or bow ties, see-through bottles containing different numbers of items, eg 3 bottles with beads, 3 in one, 10 in another and about 30 in the last.

■ language of comparison of size

P 2 Let's build a tall tower
You don't need to stick boxes together unless you want to keep the tower for a few days.

- *'Can you make a tower as tall as you with boxes? How many boxes did you use?'*
- Challenge groups of children to build the tallest tower.
- If you ask groups to work at the same time, they can also see which tower stands up longest.
- *'Which tower is tallest/shortest/strongest/ has stood up the longest?'*

Other skills: *'Which tower uses the fewest/most boxes?'*

Other contexts: make Rapunzel's tall tower, the lamp post in Narnia out of the inner tube of a carpet roll, a tent out of poles and sheets.

> **Provide:**
> a selection of cardboard boxes.

■ language of weighing, measuring

P 3 At the baby clinic

- Many children will be familiar with the baby clinic and will have watched babies being weighed. Discuss other activities that take place at the clinic, eg measuring the babies, selling baby food and clothes, making appointments.
  'Does this doll weigh more than that one?'
  'Which was the heaviest baby?'
  'Which was the oldest?'
  'What did you buy and how much did it cost?   How much did you spend altogether?'
  'How many babies did the nurse look at today?'
- Children can make appointments in the appointments book and give out appointment cards. Let children record these through emergent writing and their own representation of numbers.

Other contexts: fruit shops, supermarket, farm shop, market stall.

> **Provide:**
> balance scales and dolls, weight progress cards, appointments book, tins of baby food for sale, pennies etc.

Number

■ addition and subtraction of pence, finding totals and giving change

P 4 The jumble sale

- Ask children to write price labels for sale items. Pictures of coins next to the items can help (or stick on a coin).
- Children can take it in turns to be shopkeeper or customer. Customers can
  1) hand over coins to make the right amount;
  2) hand over 10p and be given the correct change for one item;
  3) buy more than one item and work out the total cost.

Other contexts: the travel agent (have holidays for £100, £200 and £300 only and make £100 notes or use ones from Monopoly), post office, farm shop selling fruit and vegetables, cafe (write a price list and read it with the children, eg in the cafe juice costs 50p, baked potato costs £1 and fruit salad is 20p).

> **Provide:**
> coins, paper labels and pens, items for sale, a till or money drawer

■ **early shopping experience, counting items to 10, representing numbers, tallying**

P 5 How many apples today? **IP 6**

- You could change the number of fruit on the stall each day, eg 6 apples, 5 oranges, 4 bananas, 3 plums, 2 kiwi fruit, 1 grapefruit. One day have 3 of everything. Count the fruit each day for mental maths.

- Making surveys gives children a reason to count and involves them in representing numbers in some way. Let children choose their own problem or pose one yourself, eg *'Which of these fruits do children in our class like best? How could we find out?'*

Other contexts: any shop can work in this way, eg the baby shop, the post office, the supermarket.

> **Provide:**
>
> a fruit barrow, real or pretend fruit, shopping bags, purses, coins, scales, paper and pens for shopping lists and orders, bike/cart for delivery person.

■ **language of maths, reading numbers, open for assessment**

P 6 Schools

- Set up a 'school' and include the maths table in it if you can. You can suggest that they have a maths lesson and ask children what they would like to do in that lesson. Remind children that they will need to bring what they do to the review session.

- Specific tasks you could give are writing numerals, writing numbers as far as they can go on a number line on IP 14, playing shops using IP 5.

Other contexts: set up an office, or a ticket office for buses, planes etc.

> **Provide:**
>
> paper and pencils, interesting pens, some resource sheets, an IP page and washable spirit pens, maths games if you have them, cover games from the practice books, appropriate books and equipment, eg number lines etc.

Pattern

■ **making repeating patterns**

P 7 Wallpaper shop

- Set up a printing shop to make wallpaper for the play house and doll's houses etc. Try to include some common 2D shapes, eg the end of a kitchen roll tube to make a circle.
  'Tell us about the pattern you made.'
  'What shape comes next?'

Other skills: draw out the names of shapes when discussing the patterns that have been made.

  Other contexts: make wrapping paper for presents and stamps for the post office, pastry play using different shapes of cutters, eg stars at Christmas.

> **Provide:**
>
> paint or ink pads, plastic or sponge shapes, rubber stamps, pieces of potato and other vegetables.

■ **representing numbers and using repeating pattern**

P 8 At the jewellery and hat shop
- Set up a jewellery and hat shop where children can order items to be made, eg a silver crown with 10 blue jewels, or a blue hat with a red and yellow pattern around it, or a bracelet with 6 rubies.
- Children take turns to be the shopkeeper. Customers order what they want by writing or drawing in the order book. The shopkeeper makes the jewellery or hat to fit and then decorates it – usually with the help of the customer.
  'How many blue stones?'
  'How many red stars?'
  'Tell us about the pattern on the crown.'

> **Provide:**
> glitter, coloured paper, sequins to make jewels, card, sticky tape, glue, an order book.

The maths table

What is the maths table?

The maths table is an area in the classroom where children can go, as part of a lesson or in their own time, to investigate mathematical ideas using a variety of apparatus.
The activities can be
- planned by you and explained to the whole class
- planned by children during the first part of a lesson.

The activities in this book are generally teacher-independent activities. A few sample activities are given here and there are further ideas in each reception lesson plan.

Setting up the maths table

It can be helpful to place the table near to where maths apparatus is stored already, but that is not essential. Make it clear to children what resources they may get out of cupboards, and how they are to tidy up.
Laminated mats can protect the table surface from felt tip pens.

For storage, use cloth or laminated pockets, or covered boxes so that children can take things as they need them.

What to provide
Any mathematical apparatus can be used, even those things that usually lurk at the back of cupboards!

Always provide
- a variety of pens, crayons, pencils, stamps, sticky labels, stickers (eg stars), scissors and glue, stencils (eg NCM).
- a variety of different kinds, colours and shapes of paper.
  Keeping these at the maths table can give the activities done there status and the use made of them will give you insights into children's thinking.
  Include worksheets for anything you want children to practise, eg writing numerals.
- number cards, wooden numbers, tactile numbers, dotty cards (see equipment list), spinners, dice, counters, cubes/buttons/teddies/shells etc to count, sorting toys.

- interesting containers such as hinged lid boxes, 'treasure boxes' (cover chocolate boxes with shiny paper and stick on sequins).
- number lines of different lengths and types, eg NCM resource sheets, a Unifix plastic number line that cubes fit into, small toys to move along the line.
- beads, strings, including 10-strings with 5 beads of one colour, 5 of another (see IP 14), old style bead abacus.

Other equipment that can work well

- number and size jigsaws;
- small construction toys such as Polydron;
- playdough/Plasticine, pastry cutters, tin lids, blunt knives, a variety of bun trays with various arrangements of spaces;
- Cuisenaire rods and other structured number apparatus;
- old clocks, timers, tockers, play clocks, egg timers;
- plastic 2D and 3D shapes, boxes that can be opened out;
- scales, play money, money boxes, price labels, blank labels, supermarket bills, notebook, pencil and calculators, toy phone, catalogues;
- mirrors, sticky shapes, grid papers;
- mathematical games and mats, eg from NCM.

Interactive mathematical displays

Interactive displays on walls around the table can encourage children to investigate mathematical ideas.

Vary the displays (see below for ideas). Use laminated surfaces and Blu-Tack (or Velcro) so that children can move items on the display and stick on number cards etc.

- Fix IPs to the wall and children can stick or Blu-Tack on numbers, eg counting the buns on IP 5.
- Make a washing line as on IP 15 with pegs and numbered paper shirts.
- Laminate a large picture of a birthday cake, keep candles in a pot; keep blank cards and wooden numbers nearby so that children can make greetings cards.
- Have a number line of some kind, eg large laminated number cards in pockets so that they can be moved; put small hooks beneath a wall number line so that beads can be threaded and hung by the numbers.
- Have numbers and number words in different languages and scripts in a format that allows them to be removed and copied if needed.
- Put up a laminated height chart.
- A 'number board' can be kept up all the time with pictures and photos of numbers in the environment that can be changed regularly, eg door numbers, price tags, bus numbers, car registration numbers, receipts, birthday cards, 'the number of children in our class today is...' etc.

Display books that have a mathematical theme

- number rhyme and counting books
- small books made by you or by the children that deal with particular ideas and concepts can give a focus to work at the maths table, eg 'Tina's book of big numbers', or 'Our book of shape robots' (IP 9).

Providing tasks

At the maths table keep a 'things to do' box containing laminated cards that give ideas, eg a picture of scales and parcels to weigh.

To help children with specific tasks, use laminated cards with the beginnings of instructions on them, eg 'Find out about ...' or 'Today I want you to ...', 'Play the ...' and Blu-Tack on names of activities, eg 'Play the fish number game'. Read these together during whole class sessions.

Keep lists of tasks or children's names as a 'visitor's book' so that children can tick their name when they work on the maths table, or tick the task completed. These lists give children valuable opportunities to learn to record.

Following up work at review time

Let children bring their work to review time and talk about it. Knowing that they will have to show and explain their work at this session will help to keep children on task when at the table.

Not all work done at the maths table needs to be recorded and any recording that is done need not be formal. Children's own informal recordings can give great insight into their thinking and many children will enjoy recording 'so that you won't forget.'

Keep these recordings for displays and as examples of children's progress. They also work well when put together to form a class book. For example, if everyone were to do MT 7 Draw and count, every child's picture and writing could go in a large 'floor book' for the book corner that you can share together at story time.

Annotate other work with your comments and keep it in the child's record folder to demonstrate the level the child is working at.

Evaluating the success of the maths table activities

Use what comes up at the review session time to decide what needs to be done next to develop thinking and to plan tasks for the next day or week.

Some very young children might play and scribble in a way that is not what you intended but children learn from this free play. They will also learn from what others concentrate on and talk about at review time.

If children don't respond to an activity, set up a similar one in a different context.

Sorting and matching

■ sorting and matching

MT 1 Sort it out **N/R**
- You could do CG 1 before you do this. Select a variety of sorting objects, so that there are obvious groupings, eg red and yellow cars, trucks and boats. Children can see how many different ways they can find to sort this group of objects, and report what they did to the review session.
- Move on to sorting shapes and numbers. This might be appropriate for some nursery children as well as those in reception. '*Put all the shapes with 3 straight sides here and those with 4 or more sides here.*' '*Sort all these dominoes into sets. We could put all the ones that add up to 2 in this box.*'

■ making pairs

MT 2 Finding partners **N/R**
- Put out a selection of things such as bat and ball, toothbrush and toothpaste, 2 Lego pieces that can go together etc and ask children to match them up.
- Later, match dotty cards to number cards, and match number cards to groups of creatures (eg 5 fish) on IP 8.

Number

■ more than/less than

MT 3 Ducks in a pond **N/R**
- After a whole class session on counting and using the words 'more than' and 'less than' ask, '*Make 2 ponds, one with more ducks on than the other,*' or '*Put 4 cows in a field and put more than 4 sheep.*'

Provide:
items on the table that reflect the context you will use, eg two 'ponds' and plenty of ducks, fields with cows or sheep, number cards

■ more than/less than/same number as

MT 4 Draw more whales **N/R**

- *'Draw more mice/whales/umbrellas/flowers. How many are there now?'*
  'Put one more window on each house. How many now?'
- *'Draw a flower for each butterfly, or 2 balls for each clown.'*

> **Provide:**
> IP 4 and washable spirit pens

■ counting and representing numbers

MT 5 I can count and write numbers **N/R**

- Give children a specific task, eg *'Draw Jessica's birthday cake with 4 candles'*; *'Give each ted on IP 7 one more balloon and write the number of balloons on the ted's tummy.'*

> **Provide:**
> IP 7, tactile numbers, other number cards, items to count (eg birthday cake candles or 'five speckled frogs' or buttons), pens, paper etc

■ counting to 5 and writing numbers

MT 6 Dots and stickers **N/R**

- Let children make groups of up to 5 dots, sticky pictures, stamped pictures etc, and then write the appropriate number beside each group. These can be displayed on a 'counting we did today' board.

> **Provide:**
> paper and stickers, sticky dots, rubber stamps

MT 7 Draw and count **N/R**

Give specific tasks, eg draw:

- the numbers 1, 2, 3 and the right number of cherries next to the number;
- 2 birds, 2 cats, 2 stick people;
- 5 clowns and a ball each;
- 3 mice and find a number 3 card to go with them.

> **Provide:**
> number cards

■ counting and doubling

MT 8 Spotty butterflies **N/R**

1. Give each child a number card.
2. *'Make a butterfly with that number of spots on each wing.'*
3. Once children have made their butterfly, they work out how many spots are on it altogether and stick it next to that number on a number line.

■ counting and one-to-one correspondence

MT 9 Birds and worms **N/R**

- Children take a number card and put out that number of birds. They make a worm for each bird. At review time emphasise the number of birds matching the number of worms. '*Harry has 4 birds and 1, 2, 3, 4 worms.*'

> **Provide:**
> number cards, playdough, plastic birds or pictures

Calculating

■ language of number

MT 10 Number stories **R**

- '*Take a number card. Put that many sheep in a field. Move some to the barn. How many did you move? How many are left?*'
- Encourage children to make up number stories in other contexts, eg '*Three candles were blown out, how many are still alight?*' '*3 birds and 2 more,*' etc.

> **Provide:**
> IP 11, plastic or cotton wool sheep, fields, number cards, birthday cakes

■ solving simple number problems

MT 11 Dominoes **R**

- Take some dominoes out of the set if you want so totals don't go beyond a certain number, eg 6. Put out numbered quoits or pots. '*Work out how many dots altogether. Put the domino into the pot with that number on the front. Which group do you think will have the most?*'

At review time ask, '*Did it work out as you expected?*'

> **Provide:**
> dominoes, quoits or pots with numbers on them

■ language of ways of calculating .

MT 12 How did you work it out? **R**

- Once you have introduced the think clouds on IP 10, keep these materials out for children to use for recording their mental strategies during number lessons. This could be after a session with a number rhyme book. Encourage children to draw cubes and fingers etc. At this stage you are aiming to encourage children to talk about what they do and 'different methods' are likely to involve different equipment.

> **Provide:**
> coloured cut out think clouds (as on IP 10), number cards (tactile ones if you can), pens, sorting toys (cubes/buttons, coins, carriages on a train, play people, shells, acorns), trays etc

■ sharing numbers out equally

MT 13 Share it out **N/R**
- *'Share out the buns so that it is fair. Draw the buns that each ted gets.'* Let confident children try large numbers of items.

Provide:

items to represent buns etc and toys that want an equal share, eg 3 toys and 9 buns

■ solving problems of sharing and fractions

MT 14 Cut it up **N/R**
- Children work in pairs and imagine that they are at a party. Let them use a blunt knife to halve items to share between them, eg a pizza, a sandwich, a bun. They can bring their work to the review session.

Provide:

playdough birthday cake, sandwiches, pizzas etc, real fruit etc

Measures

■ comparing measures

MT 15 Make it longer **N/R**
- *'Draw longer whiskers on the cats and mice. Make the big cat's whiskers very long.'*
- *'Draw in a long snake. Then draw a longer snake.'*

Provide:

IP 4

MT 16 Who is the tallest? **N/R**
- During the day, pairs of children should mark each other's height and name on the chart. At the end of the day discuss who is the tallest 3 year old, the shortest 4 year old, etc.

Provide:

a laminated height chart, pens

MT 17 What is the heaviest? **N/R**
- *'What is the heaviest thing in the basket? Show us how you know that?'*
- *'How many teds balance the doll?'* *'Do the doll and the puppet together weigh more or less than the engine?'*

Provide:

a basket of items, balance scales and plastic teds for weighing

Nursery lesson plans

These 'lessons' are short activities that last between ten and fifteen minutes and are aimed at nursery or young reception children. They are each linked to an Interactive Picture and are arranged in order of the IPs. Activities in the mental maths and circle games section would also make suitable 'lessons' for children of this age.

You can do the activities with a small or large group of children. Unless otherwise specified, it is assumed that the children are sitting near to you, usually in a circle on the carpet, where you can maintain eye-contact with all of them.

Use the planning grids at the back of the teacher's book to find the lesson with the learning objective you wish to focus on.

Assessment activities for nursery

Working with a small group of children on any of the activities listed here will give you many opportunities to assess the levels of children's understanding. You can also set individuals specific tasks which will show what they can achieve, eg

- *Sam, tomorrow I'm going to ask you to count the buns on the picture with me just like you did today.*
- *Rob, please show me again how clever you are at knowing how many fingers you have on one hand.*
- *Kathy, when your dad comes you can show him how you count up to 5, and how to write a 5.*

It is important not to make these tasks threatening. They help to keep children aware that there are things to learn and that you are there to help them to do that.

IP 1: How many?

Learning objectives:

- reciting numbers to 10
- counting and recognising numbers to 3/5/10
- conservation of number
- last number of count is number in set
- using counting to solve practical number problems
- writing numbers to 3/5/10

1a. One, two, three

- counting and recognising numbers to 3
- conservation of number

You need: small objects to count, eg plastic sorting toys

1. Give each child a handful of small toys.
    '*Count out 3 toys. I'll do it too. Put them in a line like this and let's count them.*' Encourage children to move objects as they count.
    '*Now put them in a small group like this. How many now?*'
    '*Hide one. How many now?*'
2. Write the numbers 1, 2 and 3 in the cards held by the characters at the bottom of the IP.
    '*Hold up 1/2/3 fingers.*'
    Point to one of the numbers you have written.
    '*What number is this? Hold up that many fingers.*'
3. Give number cards to some children who hold them up at appropriate times as you sing number rhymes, eg 'One, two, …' *Seven Dizzy Dragons*, page 2.

1b. Caterpillars

- conservation of number
- reciting numbers to 10

You need: 10 paper circles or quoits

1. '*Look at the caterpillars on the IP. How many sections does each one have? Let's count.*' Both caterpillars have 10 sections, but one is more spread out so is longer than the other.
2. '*Let's make an enormous caterpillar right across the carpet with 10 sections.*' Use paper circles or quoits. '*Now let's make a shorter caterpillar but still with 10 sections.*'
3. '*How many fingers on both hands? Spread them out. Is it still 10? Squash them up small. Do you still have ten?*'
4. Keep reciting the sequence 1–10.

1c. How many?

- counting to 3/5/10
- last number of count is number in set

1. Count objects on the page. Ask children to mark things counted, eg by drawing a line around the set of 3 bears and writing 3 next to it.

2. *'How can we make sure that we don't count something twice?'*

3. As you count the items say all the numbers *'One, two, three, there are three in this set.'* Emphasise that the last number you said is the number of items in that set.
Look out for children who count to three then say there are four objects. *'We said 1, 2, 3 and so we know we have 3 bears here. The last number we say when we are counting is the number in the group. Let's count them together again to see that we have 3 bears here.'*

These are some of the items on the page:
1 ark with 10 sets of 2 animals arranged in 2s
2 caterpillars, one shorter than the other, each with
   10 sections including the head
3 bears of different sizes
4 rabbits
5 monsters
6 bikes and 6 ladybirds
7 dizzy dragons
8 cats
9 birds
10 astronauts

1e. Handwriting practice

- writing numbers to 3/5/10

1. Use the top border of the IP to show the correct formation of numbers. In the handwriting book numbers are sometimes practised in groups depending on which direction we start writing them. During handwriting time, practise writing numbers and letters that start in the same direction in your schools' handwriting style, eg 4 starts down as do b and h.

1d Cups of coffee

- using counting to solve practical number problems

You need: everyday objects to count

1. Do some counting in everyday contexts, eg the number of cups of coffee needed for all the teachers, the number of eggs/apples you bought yesterday, the number of pound coins in your purse, the number of children allowed in the playhouse at any one time.
'Did I buy enough apples for everyone in this group to have one?'
'Is there a biscuit in the packet for everyone in the class?'

2. Put up as many classroom notices as you can for children to learn to read and use numbers, eg make a list 1–10 for children to sign to take turns on the new tricycle, wear the police helmet, use the computer etc.

IP 2: **Games to play**

Learning objectives:

- using the language of size
- conservation of number
- recognising and ordering numbers to 3
- counting dots to 6 and matching to number
- one-to-one correspondence

2a. Three butterflies

- counting to 3/5
- conservation of number emphasising not counting items twice

You need: small sorting toys or bricks

1. Count the butterflies on the IP. Give children small sorting toys so that they can count out the same number of items, eg 3 bricks, one for each of the butterflies.
2. Children sit with 3 bricks in front of them. Show how to move them one at a time as you count 1, 2, 3, very slowly. State clearly, '*You must be very careful not to count any brick twice. Check by counting again if you think you might have counted something twice.*'
3. '*Hide one/two behind you. How many in front of you?*' '*Now hide all of them. How many in front of you now?*'
   (Use the word zero right from the start but link this to 'none' and 'nought'.)
4. Go back to counting all 3 one at a time and moving them. Help children to say just one number word for each brick and remind them that the last number said is the number in the set. '*1, 2, 3; there are 3 bricks altogether.*'
5. '*Put the bricks you've just counted in a heap close together. How many now?*' '*Now spread them out. How many?*' (Remind them that the number doesn't change.)
6. Count the butterflies again and write 1, 2, 3 beside them. Repeat the words 'one, two, three,' until all children can say them by rote.

It will take time for some children to be able to say one number word for each item.

Follow on

7. Link this to taking 1, 2, 3 steps on a large floor number track and to number cards 1, 2, 3.
8. Count the rabbits and write numbers 1, 2, 3, 4.

2b. Children 1, 2, 3

- recognising and ordering numbers to 3/5

You need: a box or feely bag containing wooden numbers or cards 1–3 (later 5), washing line

1. Show number cards 1, 2, 3 and demonstrate putting them in order. Make a washing line with 1, 2, 3 (see CG 16).
2. Pass around the box. In turn, children take a number out of the box, hold it up and everyone says the number.
3. When children are good at this, give three children a number each and put them in order by matching them to the washing line.

Follow on

4. Develop this game with 4, 5, then higher numbers and with counting out cubes to match the number cards.

2c. Teddy tummy numbers

- matching numbers and dots to 6   *ladybirds*

You need: counting toys, number cards 1–6 and dotty cards 1–6

1. Count the bears slowly, pointing to each one as you do so.
   '*Count out 6 plastic teddies,*' '*Show me 6 fingers,*' '*Find 6 on the floor number track,*' etc.
2. Point to a number on a number card / a teddy with a number on his tummy. '*Who can come and pick up the dotty card that goes with that number?*'

Follow on

3. Map (draw a connecting line from) the numbers on the teddies to the 1–6 spaces on the dragon. (Children can repeat this activity on the maths table.)
4. Sing a rhyme about teddies, eg 'Ten little teddies' *Seven Dizzy Dragons*, page 8, starting just from 6 teds if you want.

2d. Teddy game

- language of size

You need: teddy spinner, counters, Blu-Tack

1. Explain what large and small mean using objects.
2. Fix a counter to the house with Blu-Tack. *'We're going to play the teddy game. When the spinner lands on the big ted, we move forward to the ted that is just bigger than the one we started on. When it lands on the small ted, we move back to the ted that is just smaller than the one we started on.'*
Children take turns to spin and move the counter. The game ends when they reach the largest ted.
3. Play the game with 2 or 3 teams with different coloured counters.

2e. Handwriting book cover game

- language of size

You need: a teddy spinner for each pair, counters

This game is like the teddy game but has 10 bears and is played by pairs of children. Even if children are not yet ready for the handwriting book they can play the game. Check at review time that children can use words such as 'bigger bear' and 'That bear is even larger,' 'That is the smallest bear.'

2f. Teddy race game

- counting, one-to-one correspondence

You need: 3 dried butter beans, each with a red dot on one side

1. Split the class into 2 teams. Blu-Tack a different-colour counter for each team to the house on the IP. Ask a child from one team to throw all the beans. Count how many land with the red side upwards. Move the team's counter that many spaces along the line of bears. Teams take turns to play. The first team to reach the 6th bear wins.

2g. Red beans game

- counting to 3, matching to number

You need: 3 dried butter beans, each with a red dot on one side; number cards 1–3

1. Pass the beans round the circle. Children throw them in turn, count how many land with the red side upwards and choose the card with that number.

Follow on

2. Develop to larger numbers using more beans.

IP 3: **Nursery rhymes**

Learning objectives:

- using the language of comparison: tall, medium, short etc.
- using the language of number comparison
- sequencing events
- language of position and movement
- reading 9 o'clock on analogue clocks
- counting to 3/5 and above

3a. How tall is the giant?

- language of comparison: tall, medium, short etc

1. Ask children to suggest how tall the giant might be in comparison with themselves and with the height of the school. Other tall items: giraffe, trees, tower is taller than the house, the four clowns are all different heights out of order, the goats are three sizes in order.

3b. Can you see five giant toes?
- counting to 3/5 and above
- number comparisons

1. Count the different items on the page and compare the numbers.
   'There is just one cow, but 2 baby swans, so are there more cows, or more baby swans?' 'Are there more baby swans or more frogs? How do you know?'
2. *'Can you see someone with 3 buttons and someone with 4?'* (the clowns)
   'Can you find something so tiny you can hardly see it?' 'Can you count the legs on the ants? Try to find out how many legs altogether. How can you remember which number you are up to?'

1 cow, robot, Incey Wincy Spider (but 2 more tiny spiders), snail, wolf, Red Riding Hood, hare, elephant, giant, tower, house, troll, bridge, stream.
2 baby swans, creatures on the see-saw, fish in the stream, squirrels,
3 cats (tiny pussy in the well), snakes round the pond, Billy goats gruff, ants
4 birds in the sky, clowns
5 frogs
8 spider legs, shoes on the clowns
12 numbers on the clock
14 trees, stars

Children can continue to find and count tiny items on the maths table.

3c. What happens next?
- sequencing events
- reading 9 o'clock analogue

1. Ask children to re-tell familiar rhymes and stories involving the characters on the page.
2. Talk about the familiar time sequences of the nursery day, eg *'We have fruit and drink time after morning play.'*
3. Talk about the time shown on Sleeping Beauty's tall tower (9 o'clock).

3d. Clowns going to the circus
- language of position and movement

1. *'Where are ...?'*
The clowns are playing along **the edge** of the road.
The snakes are **around** the pond.
The troll is **in** the river / **under** the bridge.
The goats are going **over** the bridge.
The robot is running **along** the middle of the road.
Jack and Jill are falling **down** the hill.
The cow is **over** the moon.
Red Riding Hood is going **through** the forest.
The wolf is hiding **behind** the tree.
Incey Wincy came **down** the drain pipe.
Little Miss Muffet is **on** her tuffet.

IP 4: Monsters and clowns

Learning objectives:
- counting to 3 and above
- using the language of number comparison
- identifying 3s
- using the language of comparison of measures

4a. Let's count the bears
- counting to 3 and above
- using the language of number comparison
- identifying 3s

1. *'Let's count the bears / houses / cats / clowns / whales.'* (3 of each.) *'Hold up 3 fingers, let's count them carefully. You 3 children stand up. Let's count them, pointing to each one. Let's have more than 3 children standing up. What number shall we choose?'*
2. *'Are there more flowers than butterflies?'* (Yes.) *'Are there more bears than cups?'* (No, there are the same number.)
   'Who has the most babies, monster or dragon?' (Monster has 3, dragon has 2.)
3. *'Someone on the picture is holding a number 3. Can you point to it?'* (The robot. Numbers are also on the caterpillar.)

4b. Fat cat, thin cat

- using the language of comparison of measures

1. *'Which cat is the fattest? Which is the thinnest? So which one is between fattest and thinnest?'*
   Ask similar questions for other characters and creatures on the IP:
   The whales are large, medium and small and are arranged in that size order.
   The flowers bottom left are tall, medium and short but are out of order, so are the houses.
   The tower is taller than the clown.
   The baby dragons are smaller than the mother dragon.
   Link this with putting bears or dolls in order of size.

2. *'Do you think the mice will run faster than the cats?'*

IP 5: Baker's shop

Learning objectives:

- one-to-one correspondence
- conservation of number
- counting sets and recording the number

5a. Five currant buns

- one-to-one correspondence
- conservation of number

You need: 5 objects to represent buns, 5 pennies, a basket

1. Show the five currant buns on the IP and count them slowly and carefully, pointing to each one in turn.

2. Choose five children to be the five people who will come and buy a bun. Give them each one of the pennies. Sing the rhyme:
   Five currant buns in the baker's shop,
   Big and round with sugar on the top,
   Along came (name of child) with a penny one day,
   Bought a currant bun and took it away.
   The five children act out the rhyme, standing up in turn and exchanging the penny for a bun.

3. The 5 children sit with their buns and pennies in front of them.
   'How many buns are there? How many pennies? How many children are sitting there?'
   Establish again that there are 5 of each.

4. Bunch the 5 children up close together. *'How many now?'*
   Count them and establish that although they are all bunched up together, there are still the same number of children. Repeat with the buns and the pennies.

5. Match each penny to a bun. *'There are the same number of buns as pennies.'* Spread out the buns in a long line. *'How many now?'*
   Put them in the basket. *'How many buns are there, now you can't see them?'*

5b. Count the jam tarts

- counting sets and recording the number

1. Count the items on the plates, eg 2 loaves, and demonstrate how to record a 2 next to that set. Put the IP on the maths table and let children count items and write the number beside them.

5c. Count the currants game

- counting and one-to-one correspondence

You need: a dotty spinner marked with zero, 1 currant (dot) and 2 currants, bits of Blu-Tack to represent currants

1. Split children into 3 groups. Assign a large bun on the IP to each team. Teams take turns to spin the spinner. They win that many currants and stick them on their bun.
   Play until every team has at least 5 currants.

IP 6: **In the street**

Learning objectives:

- ordinal numbers
- using the language of comparison, eg taller than
- using the language of position
- using the language of comparison of number
- counting to 10
- larger numbers

6a. In the queue

- ordinal number
- using the language of comparison and position

1. Talk about the children in the bus queue.
   'Tell me about the person who is first in the bus queue? Who is third?'
   Encourage children to use colour and description as well as more mathematical language so that they develop confidence in talking in a group. You can also take this opportunity to assess language development.
   'Which is the tallest/shortest child?'

6b. Let's count the apples

- using the language of comparison of numbers and counting to 10

1. Count the fruit.
   'Are there more red apples, or more bananas?'
   'Stand up five children. Are there enough bananas for them to have one each?' (Yes)
   'Are there enough pears for all the children in the queue to have one?' (No)
2. Count other things on the page. There are:
   2 dogs, 2 fruit barrows, 3 cats, 3 apples, 4 pears, 6 bananas, 9 plums, 10 children in the queue and another 10 on the bus, oranges in rows 1–9.

6c. Anna had a cart

- counting
- comparing numbers of objects

1. Sing/say the rhyme from *Seven Dizzy Dragons*, page 17.
2. Count the oranges in rows.
   'Are there enough oranges for all the children in the queue to have one? Let's match the oranges to the children with this pen.' Join one orange to one child with a line. *'There are some left over.'*

6d. Numbers around us

- larger numbers

1. Go on a 'number walk' either around school or beyond that. You can record numbers children see, eg on cars, houses, in shops. *'Why do you think that number is there?'* (price label, road sign etc.) *'Can you read that number?'* Numbers can be ordered on a washing line (see CG 16) and some found on the hundred square IP 15.

IP 7: **Teddies and balloons**

Learning objectives:

- counting to 10
- writing numbers to 10 in order
- addition as counting on

7a. Counting balloons

- counting items out of reach and randomly spaced
- the number of items is the last number in the count
- writing numbers to 10 in order

1. *'Let's count how many balloons this ted has. What could we do to be sure that we don't count one twice?' 'The number of balloons is the last number we say. 1, 2, 3; this bear has 3 balloons. Let's write the numbers 1, 2, 3 on the balloons.'*

2. *'Who would like to help me to show how many balloons each ted has, by writing the numbers 1 to 10 beside the right teddy?'*

7b. How many?

- counting
- addition as counting on

1. All the sets of other creatures on the page are divided into two groups.
   'Let's count the ladybirds. There are 6 here and 2 more here so how many altogether?'

IP 8: **Playing by the pond**

Learning objectives:

- counting and writing numbers 0–5/10/20
- one-to-one correspondence
- recognising numerals and ordering numbers 0–10
- ordinal numbers
- early number line work, language of count on, count back

8a. Move and count

- counting to 3/5/10

1. Play CG 8 Move and count.

8b. How many playing today?

- counting from zero to 3/5/10

1. Count the items on the page.
   'How many herons fishing today? (1) Hold up one finger.'
   'How many dragons playing chase in the sky? (2)
   (also one in the race at the bottom of the page)
   Hold up that many fingers. Let's count them carefully, 1, 2. Hold up one more finger. How many now?'

'Look very carefully at the page and see if you can find 3 of something.' (Children are likely to say 'There are 3 cats.' *'Yes, there are 3, and also some more!'*)

2. *'How many purple monsters with green feet and orange spots on the page?'* (zero)
   Let children invent other things that do not appear on the page, eg a train, an enormous ice cream bigger than our room, a space ship.

3. *'How many teddies in the race? (2) How many clowns on the page altogether?' (2)*
   'Shut your eyes and think of 2 of something. Count them, 1, 2. Open your eyes. What did you think of?'
   On the page there are: 3 rabbits, 4 butterflies, 5 fish, 6 cats, 7 birds, 8 frogs (4 in the pond and 4 on the edge), 9 dragonflies, 10 mice.

8c. Clown's number track

- recognising numerals and ordering numbers 0–10 (extension to 20)

You need: number cards 0–10

1. *'Come and point to a number on the track that you know.'* Use this activity for assessment. Many children can recognise quite large numbers.
2. Hold up a number card. *'Who can match this number to a number on the page?'*
3. Jumble up a set of 0–5 number cards. Call six children to the front and give each one a card from the set. Ask the other children to help you to order them. Extend this, using cards 0–10 when children are ready for this. See also CG 16 Washing lines.
4. **Extension**: Write the numbers 11 to 20 on the track and repeat the activity above.

8d. I can write a five

- writing numbers 0–5/10/20

1. Ask children to come and write a number they know on the page (next to the number track numbers, or at the bottom of the page). Let children copy the numbers if they need to.
2. Later you can repeat this activity on the board without the IP but with numbers on display in the classroom. *'Can you write a 10 without looking at the number frieze?'*

Follow on
3. Write 11–20 on the track. Read and act out 'Twenty slithery slippery snakes' *Seven Dizzy Dragons*, page 25.

8e. Giraffe is first

- ordinal numbers to 10

1. *'The creatures are having a race. Who is first/second/ last?'*
   'Who is in front of Rapunzel? Who is just before the wizard?'
   'Who is two in front of the dragon? Who is between the robot and Rapunzel? Is the wizard going to win the race?'

2. Make a line of children. *'Tess is first. Minal is second. Who is third?'*
3. Use tidying up time to develop this language. *'The children doing the bricks are first to be ready. Who will be next?'*

8f. Clown steps

- early number line work, language of count on, count back

You need: large floor number track, toys

1. Work with the large floor number track frequently.
2. *'Clown hops along his number track. Where is he standing in the picture? If he hops on one more, where will he be? Shut your eyes and make a picture of clown standing on number 1 on his line. Let him take one hop. Which number has he landed on?'*

8g. How many fish?

- one-to-one correspondence

You need: ponds (paper/quoits), fish (scraps of felt or cubes), number cards 0–10/20

1. Give each child a pond and some fish.
2. *'Choose a number and hold up that many fingers. Now put the same number of fish in your pond.'* Go around the circle, asking children to show fingers and fish.
3. *'Choose a number you know you can count to. It can be more than the number of your fingers. Count out that many fish and put them in your pond.'*
4. Go round the circle asking each child to count their fish out loud.

Follow on
5. *'Find the number card that matches your number of fish.'*
6. *'How many fish altogether in this pond and that pond?'*
7. Sing 'One, two, three, four, five' *Seven Dizzy Dragons*, page 2.

IP 9: **Robot world**

Learning objectives:

- language of shape
- sorting and choosing own criteria
- writing numbers

9a. I can see a circle

- using the language of properties of shape, 2D (and 3D)

You need: 2D shapes

1. *'This shape* (triangle or rectangle) *has straight edges. This one has a curved edge* (circle).*'* See CG 43.
2. Ask children to choose a shape and match it to one on the IP. *'Can you find another shape like that / with 4 straight sides / with more than 4 straight sides?'*
3. Play CG 43 Match my shape.

Follow on

4. Repeat 1 above for 3D shapes. Use the word 'faces' for shapes such as cuboids.
5. Play CG 46 What's in the bag? with 2D and 3D shapes (sphere, cone, cylinder, cube and cuboid).

9b. Making sets

- sorting and choosing own criteria

You need: 2D and 3D shapes, round pot, box

1. *'Sort all the circles into this round pot, and put all the shapes that aren't circles in here.' 'Now find a different way to sort them.'*

9c. Numbers to 20

- writing numbers 0–20

1. Write numbers on the roller coaster from 0 up to the last number children know.
2. Name the numbers. *'Let's count that many children in the class.'*
   Recite the numbers often.

Follow on

3. Each time you do the activity introduce the next number. Match the numbers on the roller coaster with numbers on a wall frieze to 20.

IP 10: **How do you do it?**

Learning objectives:

- exploring and developing mental images

10a. What is clown thinking?

- exploring and developing mental images and language

1. Draw a clear number picture in clown's think cloud, eg 2 biscuits and 2 more biscuits.
   'Clown ate 2 biscuits and then he ate 2 more biscuits. How many did he eat altogether?'
2. Rub those out and draw in 3 and 2 biscuits. Ask children to tell you the number story about the biscuits in clown's head. Let them show you with fingers, cubes, biscuits, on the number line, etc.

This activity prepares children for exploring the pictures they have in their minds when they calculate. It also helps them to start to talk about numbers.

Follow on

3. Repeat this type of activity several times as a quick starter for maths sessions, eg draw 5 stick people in a cloud: '5 *children in the group, one goes away*'.
4. Encourage children to draw their ideas on paper or on the IP and when working at the maths table.

10b. What's in your head?

- exploring and developing mental images

1. Say a number rhyme, eg 'Anna had a cart' *Seven Dizzy Dragons*, page 17. *'Anna sold 1 orange on Monday and 2 on Tuesday. How many is that altogether?'* Let children show you and talk about what they can do.
2. *'Robot worked it out by adding on her fingers'*. Draw in 1 and 2 fingers on Robot's cloud.
3. Draw in other think clouds, *'Ted adds up the oranges by shutting his eyes and making pictures of the oranges in his head.'* Draw the 2 and 1 oranges in Ted's cloud.

4. Ask children to suggest other ways to do the calculation or let them show you with apparatus. Draw in their ideas.
   This activity encourages children to think about the methods they use to perform a calculation.

Follow on

5. Repeat with other calculations.
   Link these to hops along a number track so that children begin to relate 2 and 1 more to starting on 2 and taking 1 step.
6. After a number track session, let children draw what they did on plain paper.

IP 11: Inside, outside

Learning objectives:

- counting and conservation of number
- using the language of number comparison
- writing numbers to 10
- splitting numbers, and ealry addition and subtraction

11a. Birds and butterflies

- counting
- using the language of number comparison

1. *'Look at the sheep. Let's count them (pointing), 1, 2, 3, 4, 5. Hold up 5 fingers. Let's count them. Dan, come and help me to count out 5 bricks.'*
2. Count the other items on the page.
   'Are there more birds or more butterflies?'
   'How many birds this side? How many this side? Which side has fewer?' *'Two children stand this side and three this side, like the birds. Let's count the children. Which is more, 3 or 2?'*

11b. Can you write all the numbers?

- writing numbers to 10

1. Use the top border to show children how to write numbers. Emphasise the black starting dots each day. *'We start numbers at the top. Dara, come and write a 2. Where do you put your pen to start off?'*
   This can be repeated several times a week, for just a few minutes each day and followed up by independent work at the maths table. When children are ready, introduce the handwriting book.

11c. Teds in a quoit

- splitting numbers, and early addition and subtraction

You need: 3/more plastic teddies or other small sorting toys and a quoit or piece of paper for each child

1. *'Let's count our teds. (Move them as you count.) 1, 2, 3.'*
2. *'Put 1 ted in the quoit. 1 ted inside. Count the teds outside. (Move and count.) 1, 2.'*
3. *'Can you have a different number of teds inside and outside?'*
   Use zero if children do not mention it. *'Zero inside, so all 3 teds are outside. Now put all 3 teds inside. Now there are no teds outside. Zero teds outside.'*
4. Repeat with a different number of teds.

IP 12: Teddy houses

Learning objectives:

- counting
- using the language of size and description

12a. Teddies in the border
- counting

There are 22 teddies in the border: 7 small bears, 7 medium, 8 large. 5 have hats, 6 have ice creams, 3 have bow ties. There are 22 pairs of wellington boots to count in 2s.

12b. Three sizes of bears
- using the language of size and description

1. Talk about the small, middle sized and big teds in the border. *'Can you see a big bear with a hat on?' 'Can you see a little bear with bright red boots on?' 'Start counting at the big bear here and tell me the number you get to for that little bear.'*
2. *'I'm thinking of a bear. It is big. It has a green bow tie and bobble hat.'*
   'Tell me about the bear I'm thinking of. It is the third one starting from here.'

IP 13: Mice in a hole

The work from this page is all within the reception sections.

IP 14: Number lines and beads

Learning objectives:

- counting on and back along a number track
- recognising 'teens' and larger numbers

14a. Teddies out to play
- counting on and back along a number track

You need: counting toys, large floor number track

1. Use a plastic ted to show how to hop along the track at the top of the IP.
   'Ted is standing on 1. How many hops to 3?'
   'Ted is on 6. How many steps back to zero?'
2. Repeat this activity with children standing on the floor number track. Ideally, make a track outside and put out bean bags and toys etc so that children can play games.

14b. Let's count
- recognising 'teens' and larger numbers

1. Before you start write the numbers into the beads in the border.
2. Count aloud with children as far as they can go, pointing to the numbers in the border. Children need to be able to say the number names and this will be by rote at first.

IP 15: Washing lines and hundred square

Learning objectives:

- counting to 10/100
- ordering numbers to 5/10/above
- linking 100 square with number line
- recognition of numbers

15a. How far can you count?

- counting to 10/100

You need: large floor number track

1. Gather children around a large floor number track.
   'How far can you count?' Let a confident child walk along the number track as others count. Ask *'What comes after 10?'* and if some children know some of the numbers, move to the IP and point to them. (It can be helpful to extend the number track to 15 or 20.)
2. *'Who can see a number on the IP that you know?'* Let children come and identify numbers. Circle some numbers children know, eg ages of grandparents, numbers in stories and rhymes (10 green bottles, 3 bears), number of children in class today, ages of older brothers and sisters, house numbers.
3. Point to and name numbers to 100. (It is appropriate for children to do this even if they are not confident with counting to 5.) Let children chant with you to 20/30/50/100, pointing to the numbers as you do it.

15b. What comes next?

- ordering numbers to 5/10/above

You need: number cards 0–5/10

1. Point to numbers in the top row of the 100-square and let children recite them.
2. When most can do that, move on to CG 16 or 17 (washing line circle games). Use cards 0–5 or 10, muddle them up and ask children to order them.
3. Use the T-shirts on the IP to write in numbers as needed. *'What comes next after 2?'*

15c. Monster hundred square

- linking 100 square with number line

1. Draw part of the 100-square (to 20) on the playground or lay out bits of paper in the hall. Ask children to step along it just like on a number track. Draw attention to the need to go right back in order to move from 10 to 11, just like reading words in a story.
   While some children step along the numbers, count with the others, carefully chanting one number word for each step.
2. Demonstrate how you can cut up a paper number track to 20 or 30, to make the top part of a 100-square.
3. If you have space and time, extend your floor drawing to 50/100. *'Kate, go and stand on your door number, 47.'* *'Stevie, your Gran is 71. Can you find that number?'*

15d. Patterns in numbers

- recognition of numbers

1. Help children to see the patterns in numbers, counting on from 20. Point to 21, 31, 41, 51 etc. Recite numbers from 20 to 100 several times and encourage children to join in where they can.

IP 16: **Picnic**

Learning objectives:

- sharing equally between different numbers
- language of sharing
- solving number problems

16a. Along comes another ted for tea!

- solving number problems
- language of sharing
- sharing equally between different numbers

You need: tablecloth, plate, 6 or more buns, 1 ted at the table, 2 teds hidden, a bell

1. Talk about how many buns the 1 ted can have. Just as the children have worked it out, ring the bell. *'Oh, the door bell, another ted has come for tea.'* Let children work out how many buns each ted will have now. Then ring the bell again.
   If children can do this, try again with 12 buns and up to 4 bears.

IP 1 How many?

introduces writing numerals and links this to counting objects. You can return to the page many times to do short counting activities.

Learning objectives:

- counting, naming, and writing numbers to 5/above
- counting to/from 3/5/10
- writing numbers to 3/5/10
- comparing numbers
- counting in 2s

You need:

objects to count (including 3 bears if possible), lots of scrap paper and pencils, tactile (sandpaper/felt) number cards, playdough

Key words:

number, how many?, count (up to), zero, one, two, three, … one more, what comes next?

Initial assessments:

Note who

- can recite numbers to 3 and beyond.
- can count to three and beyond with one number word for each object.
- needs help with reciting or forming numbers or with one-to-one correspondence.

Lesson 1: One, two, three

Level: W–1
Objectives: ● counting numbers to 3/5/10

Whole class starter

Time: 10–20 minutes

1. Play CG 8 Move and count to assess children's skills. *'Hold up three fingers each and let's count them together. How many altogether? What is one more than 2/3/4? How far can you count up to? Can you come and point to a 3?'*
2. Count the three characters at the bottom of the page, pointing and counting slowly. Write numbers 1, 2 and 3 on their boards. *'Who is holding the number 1?'*
3. Trace numbers 1, 2 and 3 in the air, while you give instructions: *'Two is round and down then across.'* Demonstrate tactile number cards. Ask children to write numbers on scrap paper.
4. Find other twos and threes on the page. Draw a circle around things, eg two of the rabbits and three of the monsters.
5. Ask questions about the IP relevant to the experiences of your children. *'How many monsters altogether can you see?'* *'Are there more caterpillars or more dizzy dragons?'* *'Which number comes just after two?'* *'Is 3 fewer or more than 2?'*

Group time

Teacher focus

Time: 15 minutes

1. Count a variety of objects and write numbers appropriate to the children's experiences eg put out objects in 2s or 3s, or match objects to number cards, trace over tactile number cards.

2. Keep the IP on display. Ask individuals *'What can you see?'* to get a more detailed picture of the child's knowledge of numbers and counting. Note who can say the number words in the right order independently.

Working more independently

A. **Maths table**: Put out paper and interesting pens, number cards, a number track to 10, sorting objects and trays. *'Can you write the numbers 1, 2, 3, … Put that number of counters by each number.'*

B. *'Lay the table for the 3 bears/play people using cups, plates etc.'* Emphasise that each bear needs one cup, one plate etc to encourage one-to-one correspondence. *'How many plates? How many bears? Would there be enough plates if another bear came along? How many more would we need?'*

C. **Support**: *'Make playdough buns, one for everyone in the group/class, each one with 3 currants on top.'*

D. **Extension**: *'How many can you count up to? Make a "train" of cubes, count them and bring them to review time.'*

Whole class review

Time: 10 minutes

Key idea	Counting

- Ask children to show their work. Help children to identify their threes. *'Can you count the currants on your bun for us?'* *'How many plates did you need for the three bears?'* *'Did you have more plates or more cups or the same number?'* Encourage children to ask one another questions. *'Who would like to ask Rik a question about his robots?'*

- Practise writing numbers in the air once more and demonstrate how to use the tactile numbers again.

- Note who will need support with numbers to 3 and who can be extended beyond that.

Alternative starters

1. *'Let's see if we can count up to 5 and count out 5 children. Who can see a 5 on the page?'* (Include all children in this work, even if they are still working with lower numbers.)

2. Read the Three bears / Billy goats gruff or make up 'Snow White and the 3 dwarves' or 'The 3 Dalmatians'!

3. Trace all the numbers 1–10 and recite up to 10. This will be by rote for some children but that is a necessary stage.

Lesson 2: Going for a bike ride

Level: W–1

Objectives:
- counting and comparing numbers
- counting on and back from a given number

Whole class starter

Time: 10 minutes

1. Sing a rhyme (with finger counting) that counts on or back at least to 9, eg 'One, two, three, four, five' or 'Ten little teddies' *Seven Dizzy Dragons* pp 2, 8. Count items on the page, checking that most children can count at least to nine if they are to do the PB page.
2. Ask children to count the monsters (5) and the bikes (6) on the page. *'Are there enough bikes for the monsters to go for a ride?'* (Yes and one left over.) *'What if the dizzy dragons want to go for a ride?'* (There is one too few.) *'Are there enough if the rabbits/ladybirds/astronauts want to go for a ride?'*
3. Play CG 15 How many more is your number? including the follow on and do some counting on and back.
4. *'Let's start counting at number 6 and count back to number 2.' 'Now shut your eyes and just imagine the numbers. Let's start at 8 and count all the way back to one/back as far as 4.'*

You need:

things to count, number cards, small bits of paper with 0–3/5 dots/pictures, rubber stamps, floor number track, playdough

Key words:

count on, count back, more, less, fewer, before, after

Initial assessments:

Note who

- can count out objects reliably.
- can count on and back from a given number.
- needs more help with more than/fewer than.

Group time

Teacher focus

Time: 10–20 minutes

1. Say the number names in order and take one step along the floor number track for each number word.
2. Move children to more abstract thinking by asking *'If you were standing on number 2, how many steps would you need to make to get to 4?'*

Working more independently

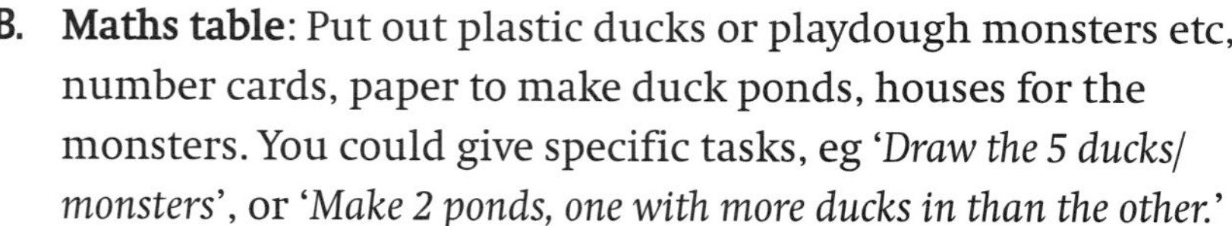

A. PB 2 p2
B. **Maths table**: Put out plastic ducks or playdough monsters etc, number cards, paper to make duck ponds, houses for the monsters. You could give specific tasks, eg *'Draw the 5 ducks/monsters'*, or *'Make 2 ponds, one with more ducks in than the other.'*
C. **Support**: Give out paper with 0–3/5 dots/pictures. Children write the appropriate number on each piece of paper. Provide stamps and paper so they can make more.
D. **Extension**: *'All the monsters and the dragons and the three bears want to go on a bike ride! How many bikes will they need altogether?'* (15) *'What if the astronauts want to come too?'* (25) *'Is that enough bikes for everyone in our class? Find the number on a number line and show us at review time.'*

Whole class review

Time: 10 minutes

| Key idea | Understand 'more' and 'less' |

- *'How many monsters in this house? Show us a picture with more dots/monsters than that. Count them. How many more?'*
- *'Is 5 more than 3? How many more?'*
- *'Count how many dots on this. Point to each dot in turn. Get that many teddies out of the box. Now get one more out. How many now?'*
- *'Let's all start at 7 and count back all the way to zero / to 3. Is 7 more than or less than 3?'*
- *'Tell me a number that is more than 2. Tell me which number is 3 more than 1. Use your fingers to help you.'*
- Note who needs more help with counting out objects (CG 8 Move and count) and who could move on to larger numbers.

Alternative starters

1. Count back from ten to zero with a rhyme eg 'Ten green bottles' or 'Rockets', *Seven Dizzy Dragons* p5, matching the words to number cards. Give out a set of zero to 10 number cards, one card to each of eleven children. Muddle them up and then ask them to get in order starting with ten. *'Is 7 teddies more or less than 5 teddies?' 'We have 7 children and 6 cups. Are there enough cups for all the children?'*

2. Play a whole class race game using the two caterpillars on the page and a spinner marked just 0–2. Number the caterpillars from 10 to 1 (on the head). Each team/pair of children starts on the number 10 (mark the position with a pen). They spin in turns and move that many spaces towards the head. They must get the exact number to win. Children can then play the snake game, on the cover of PB 3.

3. CG 6 (Count on, count back).

4. Count back on the large floor number track, eg *'Stand on 7. How many steps to 4?'*

5. Over the next weeks, extend one of your large number tracks beyond 10 and help children to count on and back taking the numbers slightly further up each time.

Lesson 3: Noah's Ark

Level: 1–2

Objective: ● counting in twos

1. Count the pairs of animals in 2s, saying '2, 4, 6, 8 etc'. Let children chant that pattern of numbers, even if they cannot count that far. There are 10 pairs of animals.

2. *'Put out the animals in 2s to go into the ark. Count how many there are.'* Count them together at review time.

3. Set out chairs in 2s to make seats in a bus. *'How many people can you get on your bus?'* 2, 4, 6, …

IP 2 Games to play

teaches children how to play games, particularly those on the covers of the PBs so that they can use them to practise mental maths skills. The notes for the handwriting book cover game are in the nursery section, Game 2e.

Learning objectives:
- matching dots to numbers
- counting on and back on a number track

You need:

cubes/counters, Blu-Tack, 1–6 dotty dice/spinner, 1–6 numeral dice/spinner, PB 1, PB 2, HW

Key words:

number names 1–10, the same number as, how many, count to

Initial assessments:

Note who

- can count dots accurately.
- needs further help with counting.
- can recognise numbers 1–6.

Lesson 1: Dragon's game and Balloon game

Level: W–1
Objective: • counting dots to 6 and matching to number

Whole class starter

Time: 15 minutes

1. Play CG 8 Move and count. Then introduce the PB 1 cover game.
2. Ask children to put out some counters in one of the patterns on the dragon. *'Where have you seen these patterns before?'* (on dice and dominoes) When several children can identify some dice patterns move on to the dragon game on the IP. Children take turns to throw the dice, count the dots, match the dice number to the pattern on the dragon and cover that number with Blu-Tack. Play goes on until all the numbers have been covered.
   'How many dots? Are you sure you are right? Who will help you to check?'
3. Use the line of 6 numbered teds on IP 2 to introduce the ted balloon game (PB 2 cover) to the whole class in a similar way. Roll a numeral dice so that children just have to match the dice number to a number on the teds and cover it with Blu-Tack.
4. When children are confident matching numbers use a dotty dice so that dots and numbers are matched.
   'Which number does that match to?'
   'Count these dots, now find the number that is the same as that.'
   'On the IP draw a line between the 6 dots and the number 6 on a ted. You have mapped the dots by drawing a line to the right number. This is 6 dots and this is the number 6.'

Group time

Teacher focus

Time: 10–20 minutes

1. Supervise pairs playing the PB 1 cover game.
2. **Support**:
   'Draw a balloon for every ted.' (Decide whether to include the teds on the spinner.)
   'Draw hats/scarves/boats for all the teds in the line without one. How many did you need to draw?' (4) *'Find the 4 number card.'*
   'Make buns for the teds. Give one to each ted. How many?' (6) *'Make 3 pies. Are there enough for all the teds? How many more do you need?'* (3)
   'Draw a flower for each butterfly. How many?' (3)
   'Draw more rabbits so that there are 5/6/7 altogether. How many more did you need to draw?'
   'Put a red cube on all the teds and a white cube on all the rabbits. Make a red cube train and a white one. Which train is longer?' (red)
   'Using small sorting toys, give each bear a toy. How many? I have 4 dogs in my hand. Is there one for each ted? How many more do I need?'

Working more independently

A. Make a cube tower to match the numbers on each ted on the HW book cover (1–10)
B. Lay the IP down flat so that children can play the games on it.
C. **Maths table**: Use small sorting toys to put in trays and match with appropriate small-number cards. Take them to review time for checking.

Whole class review

Time: 10 minutes

Key idea	number recognition

- Ask children to demonstrate how they played their games.
- Repeat the matching of dots on the dragon to numbers on the teds and on number cards. Hold up number cards: *'What is this number? Show me that number of fingers.'*
- Plan with the children where you will store the equipment they need to play the games and the classroom management strategies you want children to follow so that playing the cover games can become part of your store of teacher-independent activities.

Alternative starters

1. Play the dragon game on the IP with the class divided into two teams, the reds and the blues. Play the game as before and the winning team is the one with the most of its colour counters on the dragon at the end. Add more dotty patterns on the dragon to make a longer game. These can be either different arrangements of the numbers, eg 5 could be a line of 5, or larger numbers of dots.
2. Demonstrate 'count and write the number' by writing numbers next to dice patterns on the dragon. Children can then complete PB 1 p2 or one of the HW 'count and write' pages (7, 9, 10, 11, 12, 13).
3. Demonstrate mapping by drawing lines from the pattern of 6 dots to the number 6 on a ted. Children can then do the lower part of PB 1 p3.

Alternative starters (cont.)

4. Play CG 15 How many more is your number? and demonstrate joining numbers in order on the teds on the IP. Children can then do PB 1 p5.

5. Discuss the dice patterns on the dragon before children complete PB 1 p13.

Lesson 2: Playing race games

Level: W–1
Objective: ● counting on and back on a number track

Whole class starter

Time: 15 minutes

1. Use the spinner or cube. (If you are playing on the 6 teds, only have count on 1 and 2, and count back 1, otherwise the game will finish too quickly.)

2. Show children how to play a simple race game, starting either on the ted's house or on the zero on the snake. Move on or back as given by the spinner. Encourage a variety of language such as 'count on 2 steps' and 'hop back 1' etc. Link this to a large floor number track.

3. The snake game introduces counting on and back on a number track.
   'What number comes next?' 'What number is just before/after 6?' 'Tell me a number that comes between 4 and 11.' 'What is one more than 9?'

You need:
a 'count on 1/2/3, count back 1' spinner, or a large cube labelled in this way, counters

Key words:
number names 1–15, count on one from, count back, count forwards/backwards, one more/less

Group time

Time: 15 minutes

It would be appropriate for all children in pairs to be playing some kind of race game (PB 3 cover, a game using a large floor number track, Snakes and ladders just to 25, Ludo etc.) while you assess each pair.

Whole class review

Time: 10 minutes

Key idea	Counting on and back along a number line

- If several children were having problems during group time, play the snake game on the IP again.
- Have a mental maths re-cap of the maths involved.
  'Shut your eyes and imagine that you are standing on number 4 on a large snake game. Think about where number 5 is. How many steps to get to number 5?'
  'Now imagine you are on 5. Think about moving to 3. Will you need to step further towards the snake's head, or step back towards the zero on its tail?'

Lesson 3: Names and numbers

Level: 1

Objective:
- counting, matching and mapping to 10

Whole class starter

Time: 10–15 minutes

1. Play CG 14 Find a partner.
2. Draw domino dots 7–10 on or near the IP dragon. Write the numbers 1–10 down the left-hand side of the page out of order. Blu-Tack number name cards on or near the IP. Demonstrate matching and mapping of numbers of dots, numerals and words.

You need:
number cards 1–10, dotty cards 1–10 in standard patterns, PB 2

Key words:
number names 1–10

Group time

Time: 15 minutes
Children work on PB 2 p 3.

IP 3 Nursery rhymes

gives several lessons to explore different aspects of measures.

Learning objectives:

- comparison of objects using measures
- position words and language of movement

Lesson 1: What's in the basket?

Level: W–1

Objective: ● use of language of comparison of mass

Whole class starter

Time: 10–20 minutes

1. Start a discussion about the page and the things that they can see that might be heavy, eg an elephant or a giant. *'What might be very light?'* (ant, flower)
2. Focus on Red Riding Hood's basket, what might be in it and how heavy it might be.
3. Ask children to choose objects from the middle of the circle that could fill one of the baskets but would still be light enough to carry, eg dried foods or polystyrene. Put other objects, eg 500 g mass, in the second basket to make it heavy, but not full. Talk about size and mass. Some things can be small and heavy, others large and light.
4. Start a discussion: *'Could you lift up an elephant, ship or a car?'*

You need:
a variety of everyday objects that can illustrate the words you want to emphasise, eg two shopping baskets / carrier bags, some heavy cans of food, light packets, a (heavy) large plastic water container full of water, an empty box or piece of polystyrene that is large but light, 500 g masses, a brick

Key words:
tiny, small, middle-size, large, heavier/lighter than, weighs, balances, compare

Initial assessments:

Note who

- can use some words for comparison, eg heavier/lighter than.

Group time

Teacher focus

Time: 10 minutes

1. Using a variety of objects, demonstrate to children how to weigh two things on the balance scale, encouraging children to talk about what they see and to use the language of comparison, eg *'Ted is lighter than the car.'*

2. Ask for guesses about which is the lightest/heaviest thing and ask children to place about 3 of them in order, letting all the children feel two of the objects in their hands at the same time and talking about what they can feel.

3. Children work together to produce a line of objects ordered by mass to show the others at review time. Some groups might manage to work with more than three items so have a few more objects available for children to use if they have time. Ask them to think of a way of remembering their order. Provide paper or sticky notes and pencils.

Working more independently

A. **Maths table**: Provide shopping bags, heavy and light items, balance scales, play money, a ruler and a brick to make a see-saw, small toys for children to weigh.

B. **Support**: Children put items in three carrier bags so that one is light, one is heavy and one is in-between. Make light items large (wrapped polystyrene, bags filled with crumpled newspaper) so bags are full, but light.

C. **Extension**: *'What do you think is the lightest thing in the room? What might be the heaviest?' 'How do people move very heavy things? If you had to move something very heavy from one end of the room to the other, what could you do?'*

Whole class review

Time: 10 minutes

Key idea	Comparing and ordering measures

- The teacher focus group show their ordered row of items and check them on a balance scale. Focus on their language, eg heaviest, heavier, middle, lighter, lightest. Encourage other children to question them about their work, eg *'How did you find out about ...?' 'How do you know that ...?'*

- Discuss whether a full bag is always heavy. (It depends what is in it.) It is likely that there will be misconceptions about aspects of measuring, eg that big things are heavier than small ones. These will be noticeable when you listen to children, and see what they draw and write about their activities.

- *'Tell me what you learnt today about weighing things.'*

Lesson 2: Longer or shorter?

Level: W–1
Objective: ● use of language of length

<table>
<tr><td>

You need:
ribbons of different lengths and colours, short and tall, long, medium and short toys (eg elephant and giraffe), Multilink

</td></tr>
</table>

Whole class starter

Time: 10 minutes

<table>
<tr><td>

Key words:
long, tall, medium, short, high, low,

</td></tr>
</table>

1. Play CG 41 Ribbons, to introduce a variety of words associated with length.
2. Look at the IP. *'Which snake is longest/ shortest?'* *'Which clown is tallest/shortest?'* *'Which has the longer neck, a giraffe or an elephant?'* *'What other animals can you see in the picture with long/short necks?'*

<table>
<tr><td>

Initial assessments:
Note who

● can use words connected with length confidently.

</td></tr>
</table>

Group time

Teacher focus

Time: 10 minutes

1. Use a variety of objects, to demonstrate how to compare the length/height of two objects. Encourage children to talk about what you are doing, using the language of comparison, eg *'Ted is taller than the frog.'* *'This string is longer than the pencil.'* *'This ball is bigger than that one.'*
2. Ask 3 children to make sticks of Multilink. After a short time ask them to estimate whose is the shortest and whose the longest, then put them side by side and arrange them in order from shortest to longest.

Working more independently

A. PB 1 p 4.
B. **Maths table**: Include items of various sizes, eg compare bears, Russian dolls, and use playdough to make different sized bowls, buns etc.

Whole class review

Time: 10 minutes

Key idea	Comparing and ordering length

● The teacher focus group could show their ordered sticks.
● Ask children to explain how they measure to find which object is longest. *'I am standing on a chair looking over the hedge. Am I taller than the hedge?'*

<table>
<tr><td>

Alternative starters

1. If children are going to do mapping on PB 1 p 4, start a discussion about sizes of play house cups etc. *'Let's make a set of things for baby bear. How do you know that is for baby bear?'* Show how to map and join related items on the IP.
2. Wide and narrow: the roads (also curved and straight).

</td></tr>
</table>

Lesson 3: Capacity

Level: W–1
Objective: • use of language of capacity

Whole class starter

Time: 10 minutes

1. Pass round the containers so that every child handles them. Set up a story context such as a very thirsty dragon who wants a very big drink. '*Which bottle will hold the most, do you think?*' Let children talk about this. Try to repeat what children say, using mathematical vocabulary, eg '*Tim says this bottle holds the most because it is the tallest. But Dawn says that the blue bowl will hold the most. How could we find out?*'

You need:
labelled containers of different sizes, eg jugs, tall, middle and short bottles, lots of plastic mugs, all the same size, water or sand tray

Key words:
big, small, compare, full, empty, half full

Initial assessments:
Note who
• can use words connected with capacity confidently.

Group time

Teacher focus

Time: 10 minutes

1. Demonstrate different ways to compare the capacity of two or three containers (counting how many cups fill each one or filling one and emptying it into another to find which of the two holds most).
2. Children work together to find which of two containers holds the most.

Working more independently

A. One group can explore the containers. Ask them to report to the whole class at review time, using language such as: '*This holds more than that one.*' and '*Bottle A holds 4 cups, bottle B holds 5 cups, so bottle B holds most.*'
B. Make simple graphs with plastic cups to show how much different containers hold.
C. Compare how many cups of tea there are in two (or more) of Granny's teapots, or compare Jack and Jill's bucket capacity with Little Miss Muffet's bowl.

Lesson 4: I'm going for a walk

Level: 1–2

Objective:
- use of language of movement and direction

Whole class starter

Time: 10 minutes

1. *'Look at Little Red Riding Hood going through the woods to visit her Grandma. Tell me about the way the path goes?'*
2. *'Let's make a pretend wood on the carpet. Let's use these cubes for the flowers and trees and these bricks to build a pathway. Suzie, make the doll walk along the path and we will tell you which way to go.'*
3. Encourage children to turn their head and body so that they are facing in the direction that the character is moving. Identifying the movement of an object from the point of view of the object is hard at this stage but it is a crucial skill for working with Logo.
4. At each junction on the pathway, tell children to face in the same direction and decide if it is a left or right turn by holding up left or right hands.

You need:

Multilink and some flat wooden bricks or pieces of scrap card, empty boxes

Key words:

directions, turn left/right, up, down, forwards, backwards, across, along, through, to, from, towards, away from, through

Initial assessments:

Note who

- seems confident with which words.
- finds left and right difficult to identify.

Group time

Teacher focus

Time: 10 minutes

Depending on the experiences and age of your group select some of the harder words of movement such as left and right and make up a maths story of a journey (eg the three billy goats going for a walk). You could make a route in a square, ie forward, then a quarter turn right, repeated four times so they get back where they started. (See 'Square dance' *Seven Dizzy Dragons* p 29.)

Ask the children to act it out and/or draw a picture for review time.

Working more independently

A. **Maths table:** Use pegs and large peg boards or other equipment to make routes for small toys to move along.
B. Leave the IP out for one group so that they can draw routes and make up a story about one of the characters going on a picnic.
C. **Support:** *'Make a track in the sand or on the floor and drive the tractor along it so you can tell us about it later.'*
D. **Extension:** Let children explore what the floor robot can do.

Whole class review

Time: 10 minutes

Key idea	Use of language of movement.

- While groups explain what they did, draw out the language by making clear statements, eg '*So on your sand tray you made a path so that the cars turned to the right, went up the hill and then turned left and came down the other side.*'
- Sing 'Square dance' together.

Alternative starters

1. Choose some of the words you want to focus on, eg through, turn, and forward, and use a toy to demonstrate what these words mean.
2. Choose a character on the page and trace their route to another part of the park, talking about the movements (forwards, backwards, turning etc).
3. Set out a few boxes within the circle of children. '*Drive the floor robot to Jane then over to Dan without hitting any of the boxes.*' Talk about the instructions used then make this one of the teacher-independent activities.
4. Lay out a maze or route in the classroom with tables and chairs and get children to direct others across the classroom, eg '*Take one step forward and then turn ...*' You will need to provide plenty of support about left, right, quarter/half turns and review that over the weeks in PE lessons. A red dot on the right hand for everyone can help the many children who have orientation problems.
5. Ask confident children to shut their eyes and describe to the rest of the group a route around school, eg how to get from our room to the big hall. A follow up activity for this could be to let children try to draw maps as a maths table activity.
6. At first you can just say '*turn left*' but introduce a '*quarter turn left*' and then follow that up in PE and with the floor robot.

IP 4 Monsters and clowns

develops the ideas of counting.

- counting to 3 and above
- language of number comparisons

You need:
cubes/teddies etc., number cards, play house items

Key words:
numbers to 3, same number as, less than, more than, most, least

Initial assessments:

Note who

- can count out 3 objects.
- can use the language of less/more than, same number as.
- can recognise numbers to 3 and above.

Lesson 1: The three bears

Level: W–1
Objective: ● use of language of number comparisons

Whole class starter

Time: about 10 minutes

1. Seat the children in pairs round the circle. Have ready 3 bowls/cups etc. and count them together. *'Hold up 3 fingers. Let's count them slowly and carefully.'*

2. Pass around the pot of teddies and ask each child to count out 3. *'How many teddies have you, Rosie?' 'Does everyone have the same number of teddies? I'm giving Tom another ted. Does everyone have the same number now?' 'Jill, give me one of your teddies. Who has less than 3 teddies now?'*
Continue like this, giving children different numbers. *'Who has the most/fewest teddies?' 'Who has the same number as Billy?' 'Look at your partner's teddies. Tell me who has the most?'*

3. Put aside the teddies and count items on the page. Ask individuals to come and count, then write 1, 2, 3 on the cats, butterflies, bears etc.

4. *'Shut your eyes and think of three. What do you see?' 'Think of a number 3 on a number line.' 'Write a number 3.'*

Group time

Teacher focus

Time: about 10 minutes

1. Continue activities based around counting to 3 and above. *'Put out more than 3 cubes. Choose a group with more than 3 cubes and find the right number card to go with your group.'* Encourage children to move objects as they count.

2. *'How do you know how many you have got?'* *'Which has more?'* *'Point to the group that has fewest.'* *'How many more does this set have than that set?'* *'If I add one more to (take one away from) this group, how many will there be?'* Stop children while they are counting and say *'How many are you up to?'* Then at the end clarify that the last number name we use is the number in the set.

Working more independently

A. Sort items into sets of 3 'go togethers' (on trays that can be taken to review), eg 3 blue triangles, 3 middle-sized bears, etc. Let children choose and explain their own criteria.

B. On the IP find and mark other sets of 3. (There are small items: 3 ladybirds, 3 snails, 3 three-leafed clovers, 3 flowers in the window of the castle, 3 fish; and larger, 3 beach umbrellas, 3 whales, 3 windows on each of 3 houses, 3 flags split into 3, 3 clowns, one on a tricycle, 3 spots on each wing of 3 butterflies, etc.)

C. **Maths table**: Provide playdough to make sets of 3 bowls, buns etc. Number cards 0–5 or 10, several number 3 cards, tactile number cards can all be used to mark sets.

Whole class review

Time: 10 minutes

Key idea	Compare the size of sets

- *'Who has the least number of bears?'*
  'How do you find the number of bears in the set?'
  'What is 1 more than 2?'
  'Show us your "go togethers".'

Alternative starter

1. Each child comes to the circle with 3 things that go together. Count them, count fingers, write numbers 1, 2, 3. *'One, two, three. The last number we say tells us how many. There are 3 altogether.'*

Lesson 2: Three is more than two

Level: 1–2
Objectives:
- language of comparing numbers
- ordering numbers, 1–5
- recording numbers

Whole class starter

Time: 10 minutes

1. Play CG 16 Washing lines.
2. *'Three is more than two. Hold up 2 fingers. How many?' 'Hold up one more. How many? Did you need to count them that time?'*
3. *'Look at 4 on the washing line. Is it before or after 2? What comes before 2? What comes after 4?'*
4. Say a counting rhyme up to at least 5, eg 'Five little astronauts' *Seven Dizzy Dragons* p 3.
5. *'Let's draw in another ladybird, mouse, cat, house so now there are 4. Let's put a line all around the group and write 4. Write 4 in the air.'* *'Can you see another set of 4 on the page?'* (Mother monster and her 3 babies, dragon and cats have 4 legs.)

Group time

Teacher focus

Time: about 15 minutes

- *'Let's find some tiny things on the page and count them.'* (There are 3 buttons on the blue blobby creature, 4 buttons on the yellow clown, 3 fish, 3 spots on each ladybird.)

Working more independently

A. PB 1 p 6
B. Play the following games where children have a reason to count and record their score in their own way. *'Use paper and pencil so you can remember your score and tell us at review time.'*
- *'You have 3 bean bags each. Try to throw the bean bags into the box with a monster painted on the front. Score one point for each bag in the box.'*
- Make a round target on the floor with 3 sections (either concentric circles or sections of a circle). Children throw 3 bean bags onto the target and score 1, 2 or 3 according to which section each lands in.
Expect a variety of children's own recordings of number, eg tallying, dots, numerals.
C. **Maths table:** Supply cut-out paper monsters or snakes, sticky dots or pens, and number cards. *'Make spots on your snake, count them and find the matching number card.'* Make a display of your spotty monsters and snakes with related number cards.

D. Number cards (several sets 1–5, some up to 10 or 20), small squares of paper, rubber stamps. *'Stamp and write the number.'* Encourage children to use just one stamp on each bit of paper, eg bunnies. Make a display: 'our sets of 4' etc. Leave items by the display for further work.

E. *'Draw more tiny things, eg ants, on the IP and we will try to find them and count them at review time.'*

F. **Support**: *'Draw one more mouse, cat, bear etc. and write 4 beside them.'*
'Put a red cube on each bear, a blue cube on all the monsters and a yellow cube on all the cats. Now let's count your cubes.'

G. **Extension**: *'Look at the clover. Each leaf has 3 parts: 3 here, 3 here and 3 more here. How many parts on all three leaves?'*

Whole class review

Time: about 15 minutes

Key idea	Compare the size of sets

- Talk about children's scores. *'So you know that you scored all these points. I'll count them with you… Alan scored 12. Let's find that number on the 100 square. Dani scored 13. Is that more than 12 or less?'*
- Write numbers in the space at the bottom of the IP.
- Use PB 1 pp 6 and 7 to praise the writing of appropriate numbers next to sets.
- Ask children to explain their systems for recording the scores in the two bean bag games.

Alternative starters

1. Say a rhyme with 5 in it, eg 'Five currant buns', or 'One, two, …' *Seven Dizzy Dragons*, page 2.

2. Draw in 2 more cats on the IP, then draw around the set, taking care to exclude other objects. Ask children if the set has 5 cats in it. Invite a child to come and write 5. Children can complete PB 1 p 7.

IP 5 The baker's shop

reinforces counting as a preliminary to developing the ideas of addition and subtraction.

Learning objectives:

- counting, naming, and writing numbers to 5/above
- splitting numbers and early experience of addition and subtraction
- early experience of money
- problem solving using numbers and money

Lesson 1: Five currant buns in the baker's shop

Level: W–1
Objectives:
- counting, naming and writing numbers to 5/above
- one-to-one correspondence and conservation of number

Whole class starter

Time: 10 minutes

1. Count things on the IP and write how many there are beside them, eg write 4 by the 4 jam tarts.
2. Ask children who need the practice to count out 5 pennies, 5 buns and to find the number 5 card. *'Today you will be learning about counting to 5. Who can count to 5? Let's do it together.'* Count out the five pennies slowly and deliberately, moving each one as you count. *'1, 2, 3, 4, 5. Everyone hold up five fingers. Let's count them together.'*
3. Sing the rhyme '5 currant buns' (nursery section) and practise one-to-one correspondence by giving one penny for one bun.
4. Arrange the children in groups of 4 or 5. Give each child a split paper bun and 5 currants.
   'Put 5 currants on your bun. On the picture all the patterns are different. Make a pattern with the currants that is different from everyone else's in your group. Now move your currants to make a different pattern. Do you still have 5 currants?' Establish that there are still 5 and that whatever pattern they are put in, there will always be five.
5. If you want some children to be working independently during group time, repeat the simple activity 1 above (count and write the number) so that PB 1 pages 8, 9 and 11 can be done while you work with support groups.

You need:

one pence coins, buns (bricks), currants (beads), circles of paper with line for split buns, a basket or box, 0–10 number cards

Key words:

all the number names zero to 5/10, how many?, let's count, different pattern

Initial assessments:

Note who

- can count and name numbers to 5 in order.
- can count out 5 objects.
- knows that the number of objects is constant whatever the arrangement.

Group time

Teacher focus

Time: 15 minutes

1. Repeat the activities above with children who need further work to 5. Ask each child to count out five pennies and match each penny to a bun.
2. Practise the writing and recognition of the number 5. Muddle up number cards 1–5 and ask the children to identify the five.
3. Count to five along a number track and ask each child to take steps on a desk-top number track, saying each number word as they do so.
4. Count objects on IP 5, eg the five pennies in the child's hand. Do PB 1 p 8 together. Give children some experience of counting above 5 with the items on the IP.

Working more independently

A. PB 1 p 8
B. **Maths table**: Use playdough and beads or currants to make buns. Put 1 currant on the first bun, 2 on the second, 3 on the third etc. Change the task to putting sugar on the top using tiny scraps of white card. Choose numbers for particular children so, for example, Dan puts 6 bits of sugar on each of his buns.
C. Use cutters to make 5 gingerbread biscuits with 5 buttons each.
D. Have objects to count, eg 5 teddies, 5 buns, 5 coins, a sorting tray with numbers 1–5 and either coins in it (just pennies at first) or objects to count, number cards and number jigsaws etc that include numbers to 5 or more.

Whole class review

Time: 10 minutes

Key idea	However you arrange a group of objects the number of objects stays the same.

- Let children explain their work and their counting.
- *'You have learnt a very special thing today – that 5 currants can be put in any pattern, a line, or all bunched up, but there are still just 5 currants. You need to remember this for tomorrow.'*

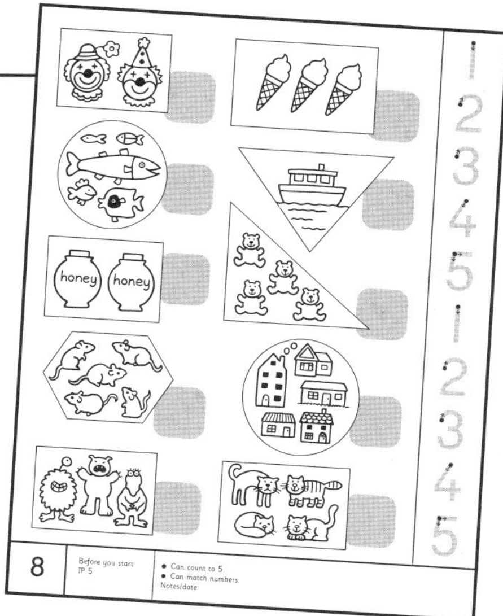

Alternative starters

1. Read other rhymes, eg 'Five little astronauts' *Seven Dizzy Dragons*, page 3. This uses the language of addition: 'one and one makes two'.
2. Repeat using a different number of buns.
3. These starters can be followed by PB 1 pp 9 and 11.

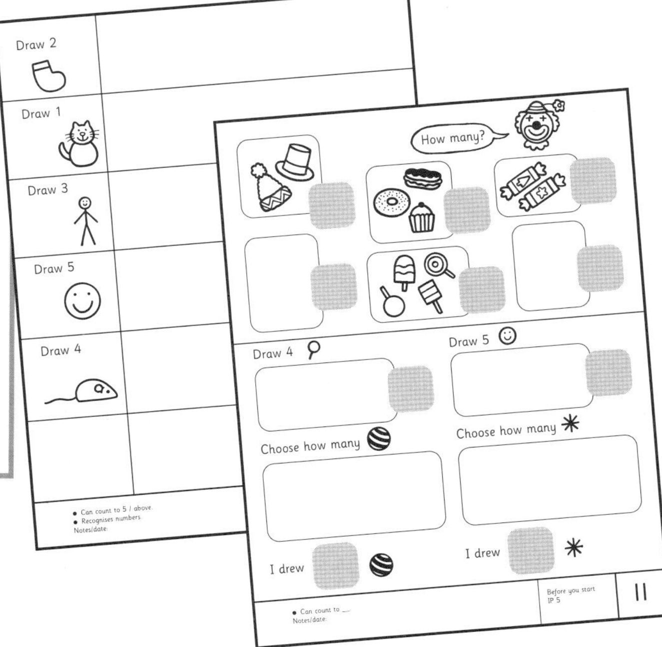

Lesson 2: Two is more than one

Level: 1–2
Objective: ● language of more/less/fewer than

Whole class starter

Time: 10 minutes

1. Make a gingerbread child with 5 buttons and compare it
    with one with only 3 buttons. *'Which is more, 3 or 5? How many
    more?'*
    *'Are there more gingerbread biscuits on the picture or more cherry
    buns?' 'How many loaves of bread?'* (2) *'Are there more loaves of
    bread, or more doughnuts?' 'Is that true that there are fewer jam
    tarts than gingerbread boys?'* (No)
2. Follow with independent activities as above and PB 1 p 12.

You need:

playdough or paper for making
gingerbread

Key words:

more, less, fewer, how many more?

- -

Lesson 3: Using pennies

Level: 1–2
Objectives: ● experience of counting with pennies
         ● early experience of coin recognition
         ● solving problems

Whole class starter

Time: 10 minutes

1. Use a bold colour pen and write price labels by each group of
    items on the IP, eg write '10p each' next to the loaves.
2. Count and buy other items on the page, eg 6 cherry buns for
    1p each, 2 jam tarts for 5p each.

Whole class review

- Ask children to identify the coins in the money
  tray and match them to real or plastic coins.
- Hold out 4p. *'Do I have enough to buy a doughnut?'*
  Hold out 10p. *'If I buy a doughnut with 10p, how
  much change will I get?'*

You need:

coins

Key words:

names for the coins, how much?
change

Group time

Working more independently

A. Set children problems to
    solve, eg *'Dad is going to the
    baker to buy 6 things for tea. Help
    him choose what to buy'* (eg 3
    doughnuts and 3 jam tarts).
    *'Laura bought a doughnut, a jam
    tart and a cherry bun. How could
    she arrange them in this long
    box? Can you find another way?'*
    *'Doughnuts are 5p each and
    Emma has four 5p coins (hold
    them out in your hand). How
    many can she buy?'*

Alternative starter:

- *'What coins would you use to make 6p?'*
  Children can then do PB 2 p 4,

Lesson 4: Splitting numbers

Level: W–1
Objective: • splitting numbers

Whole class starter

Time: 15 minutes

Play CG 29 Splitting numbers on the IP.

You need:
currants (paper/cubes), Blu-Tack

Key words:
count, split up, how many?, makes

Group time

Working more independently

Time: 15 minutes

A. Set up a baker's shop with playdough, currants etc and make different kinds of buns, eg buns with 6 currants cost 1p, cherry buns cost 2p.

B. **Support**: Practise counting, eg putting 5 beads on a string, or building a Lego train with five people on it.

C. PB 3 p 5

D. **Maths table**: Put out a variety of things with which children can make 'sums', eg paper buns to draw on currants, '5 and 2 more makes 7.'

E. Put out 2 colours of cubes or beads, paper and crayons in those 2 colours and number cards. Ask children to make different cube 'trains' that are all the same length but have two sections. *'My train has 2 red and 2 blue cubes, which is 4 cubes altogether.'* *'My train has 3 red cubes and 1 blue, and that is 4 cubes altogether.'*

Whole class review

Time: 10 minutes

Key idea	Numbers can be split in a variety of ways to give the same total.

1. Focus this session on to counting and splitting numbers, eg *'Tell me one way I could split up 5 into 2 groups?'* *'Let's count Ben's cakes. If he ate one, how many would be left?'* Check that all children know what we mean by 'splitting up numbers'.

2. Move on to the mathematical language of the 'sums' from the maths table for those children who are ready for that, eg *'How many do I need to add to 3 to make 6?'* *'Which number is 1 more/less than 4?'*

3. Play CG 28 Inside outside.

You need:

cubes, play people, number cards 0–10, hat for bus driver

Key words:

number names to 10, split, more, fewer

Initial assessments:

Note who

- can count accurately to 10.
- can split numbers, using fingers/mentally.
- can add, using fingers/ mentally.
- needs more help with counting to 10.

IP 6 In the street

has lessons which need to be repeated several times and develops the ideas from IP 5 to mental calculations to 10.

Learning objectives:

- mental addition and subtraction to 10
- coin recognition and addition of money to 20p
- solving simple problems

Lesson 1: The big red bus

Level: 1–2
Objective: ● addition, subtraction and number bonds to 10

Whole class starter

Time: 20 minutes

1. Count out 10 children, slowly pointing to each child, emphasising one number word for each child.
2. *'There are ten children in the queue. Sandy, you be the driver and send some upstairs and some downstairs.'* (Children upstairs stand up, children downstairs sit down.)
3. *'10 altogether, 3 upstairs, how many downstairs? Show that with your fingers. Are there more/fewer upstairs than downstairs? How many more/fewer?'* Let children come and count and find the right number cards.
4. Repeat this activity several times, developing the language each time and repeating the number bonds with CG 32 Make up to 10, and MM 12 Ten whispers.
5. *'Shut your eyes. Imagine 10 children. Just one goes upstairs. How many go downstairs?' 'In your head, picture 10 children and put as many as you want upstairs. How many left for downstairs? Open your eyes and tell us about your numbers.'*

Group time

Teacher focus

Time: 15 minutes

Support (counting to 10): Put a cube on each of the children in the queue on the IP. *'We want to put these children on the bus. Are there enough spaces in the bus for them all? (Yes) Move them as you count them.'*

Working more independently

Both groups need cubes/teddies/play people available.

A. Use RS 7 for children to draw their splits of ten.
B. On plain paper draw the different ways you can split ten cubes.
C. **Maths table**: Use a shoe box on its side for a bus, number cards 0–10 and 10 play people.
    'Find different ways to split up the passengers. Draw them if you want. Bring one of your ways to review time.'
D. Use an empty bus with faces stuck on with Blu-Tack and number cards to make an interactive display.
E. Draw children upstairs and downstairs in the empty bus on the IP. Leave the picture on display for a few days with relevant number cards to reinforce that number bond. Later, for example, write '5 and 5 makes 10' on the blank space on a bus.

Whole class review:

Time: 10 minutes

Key idea	Splitting numbers to 10.

- Ask children to talk about what they have done. Invite other children to ask questions.
- Early reception, read 'The big red bus' *Seven Dizzy Dragons*, p 14. Encourage using fingers and 'pictures in your mind'.
- Later reception, play MM 12 'Ten whispers'. Sometimes suggest *'Try to make a picture in your mind instead of using fingers this time.'* Children will take time to learn to do this and, for many, using fingers or cubes is essential for building up their mental images.

Lesson 2: Buying fruit

Level: 1–2
Objectives: ● coin recognition, 1p, 2p, 5p, 10p and
addition of money to 20p
● giving change to 20p (extension)

Whole class starter

Time: 10 minutes

1. Put coins in a box and pass it round the circle.
   Each child takes out a coin without looking and
   tells the others about it. The whole group tries
   to identify it. Keep the coin and pass the box on.
   When everyone has a coin say: '*Hold up all the 1
   pence / 5 pence coins etc.*'

2. Many children at this stage will be unable to
   appreciate that a five pence coin has the same
   value as 5 one pence coins. Play a simple
   shopping game, eg '*Come and buy an orange for
   5p. Do you have enough? How much change will there
   be?*'

3 **Extension**: The child at the front of the picture
   is holding 18 pence. He could buy the 3 apples
   at 4p each (6p change); 9 plums at 2p each (18p);
   6 bananas at 3p each (18p); but only 3 of the 4
   pears at 5p each.

You need:
coins, box and pretend fruit

Key words:
enough, change, total, the same
amount as

Group time

Working more independently

Maths table:

1. Put out price labels and pens. Let children
   write their own prices for the fruit barrow.
   Blu-Tack onto the IP.

2. **Extension**: '*Oranges cost 5p each. Put 5p in as
   many different ways as you can next to an
   orange,*' eg five 1p coins, two 2p coins and
   1p, and 5p.

3. '*Make rubbings of coins, bring them to review
   and we will try to tell you which coins they are.*'

Lesson 3: Fruit patterns

Level: 1–2
Objective: ● solving simple problems by copying a growing pattern

Whole class starter

Time: 10 minutes

1. Anna has the cart with oranges. Children should copy Anna's pattern of oranges with cubes. *'How many in each row? How many will be in the next row? And the next? Make the next row.'* Develop the language of number, eg more, less, fewer, one more than etc.

2. Cuisenaire rods/cubes can be used to make a 'staircase' rather like Anna's pattern, the rods lining up one end so that they go up in steps of 1 unit. At this stage, these simple patterns are giving experience with understanding what 'pattern' means. Focus on the language and on making the next row. If children can make the next row, even if they cannot say why, they are 'seeing' the pattern.

3. Let children make other patterns, eg with Polydron. *'Copy Tony's pattern.'* *'Can you start a pattern and let someone finish it?'* (See also the teddy border on IP 12.)

You need:
cubes, Cuisenaire rods and cubes

Key words:
more, less, fewer, one more than

Group time

Working more independently

Maths table:
A. Put out a variety of patterned paper, 2D shapes, stencils, coloured cubes, coloured pens, squared and other grid papers, mirrors.
B. Cardboard apple trays from fruit shops that have indented spaces for the fruit can be used to make patterns. Put oranges (cubes) on the trays.

Other lesson from this page
Ordinal number with the bus queue: *'I'm thinking of one of the children in the queue, she is wearing a pink spotty jumper. Is she 1st, or 2nd or 3rd ...?'*

IP 7 Teddies and balloons

is about telling number stories, focusing particularly on subtraction.

Learning objectives:

- subtraction by crossing out
- number bonds to 10
- doubling and counting in 2s

Lesson 1: Crossing out

Level: W–2
Objective: • subtraction by crossing out

Whole class starter

Time: about 10 minutes

1. Give each child or pair a plastic ted and 10 balloons (buttons) in a pot. *'Let's make this the third ted who has three balloons. Take out 3 balloons and leave the rest in your pot.'*
2. *'A naughty frog comes along with a big pin and pops one of Ted's balloons.'* Ask the children to hide one of the balloons behind their backs and talk about how many are left.
   'Ted had 3 balloons and now one is popped. How many are left?'
   'Is 2 more or fewer than 3? How many fewer?'
   Repeat this with different numbers of balloons.
3. *'Give your ted 5 balloons. This time the frog pops 2 of them. Put 2 buttons back in your pot. How many are left? Tell me a number story about this ted with 5 balloons and the frog popping 2 of them.'*
4. Ask children to tell you other stories about teds and balloons and build on their language, eg *'5 balloons take away 2 balloons leaves just 3 balloons. 3 is 2 less than 5. If you start with 5 and take away 2 you are left with 3.'*
5. Show children how to tell number stories by crossing out balloons on the IP.
   'This ted has 7 balloons. Cross out 3 of them. How many are left?'
   '6 mice. Cross out 2 leaves 1, 2, 3, 4 not crossed out.'
6. Invite children to pick a collection of creatures and choose how many to cross out. Let children model the crossing out with the taking away of objects.

You need:
buttons/beads and plastic teddies

Key words:
number words zero to ten, how many fewer/less than?, cross out, take away, leaves, how many left?

Initial assessments:
Note who

- can use the language of subtraction.
- can count back/forward mentally.

Group time

Teacher focus

Time: 20 minutes

1. **Support**: Assess and extend children's language with a practical activity again using buttons for balloons and plastic teds.
   'Put out the 6th ted. How many balloons does it have? Take away 3 of them. How many are left? Now put one back. How many?'
   'Now make the 7th ted and put it next to the 6th ted. How many more balloons does the 7th ted have than the 6th ted? Tell me something about the numbers 6 and 7.'

Working more independently

A. PB 3 p 9
B. **Support**: Lay IP 7 flat for children to work on.
   'Put a cube on each balloon and I will come and help you count them in a little while. Write the number of balloons on the ted.'
C. **Maths table**: Use round bits of coloured paper for balloons, a teddy stencil (NCM Stencils), glue, number cards at least to 20.
   'Draw your own ted and give it some balloons. Write down how many balloons it has.'

Whole class review

Time: about 10 minutes

Key idea	Interpreting number stories about taking away.

- As children talk about their work, extend their vocabulary by saying things back in a different way, eg if a child says *'Start with 3 balloons and pop one means there are 2 left.'* you could say *'So 2 is one less than 3. Is that right?'*
- Looking at the IP ask questions using varied language for assessment, eg *'How many more balloons would we need to give this brown teddy so that it had 5 balloons?'*
   'Which ted will be left with 3 balloons if we pop 1?'
   'What if this ted with the hat and scarf popped 2 balloons? How many would it have? What if it popped 2 more? How many would it have then?'
   'Who has more/fewer balloons, the teddy with the bow tie, or the ted with the glasses?'
- *'What have you learnt today about taking away?'*

Alternative starters

- Say subtraction rhymes, eg 5 little speckled frogs, or *Seven Dizzy Dragons* 'Ten little teddies' p8, 'Twelve red balloons' p15.
- Choose a ted, say the 10th, and work with just that ted. *'Ten balloons. Cross out 3/4/5/6. How many left?' 'What if we cross out 10?'* etc.
   'If we cross out 2, will that ted have more or fewer than the ted standing next to it?'

Lesson 2: Make up to ten

Level: W–2
Objective: ● number bonds to 10

Whole class starter

Time: 10 minutes

1. Play CG 32 Make up to 10, a counting game with ten fingers. Make the reversals of the number bonds explicit, eg '*7 and 3 go together to make ten, so 3 and 7 do the same.*'
2. Lay out large number cards in pairs, eg 5 and 5 together. '*What if we try to put these pairs in order? What could we do?*' (Keep the pairs in order on display and/or record them on a board so that they are available for later use with PBs and children will learn these number bonds.)
3. Work with the ted with 10 balloons and find ways to split up 10, eg 7 and 3. Draw a line between groups of balloons. '*How else could we split 10?*' Repeat the finger counting using a wide range of language. '*If I put 1 down, how many will be up?*' '*What do I need to add to 4 to make it up to 10?*' '*6 and how many more make 10?*'
4. Sometimes hide the number you are working with, eg '*I have 2 fingers up behind my back. How many do you need to hold up to make my 2 up to 10?*'

You need:

sets of number cards 0–10 and at least one set of large cards

Key words:

partner, pair, makes up to 10

Initial assessments:

Note who

● can talk fluently about splitting numbers.
● is beginning to recall number bonds to 10.

Group time

Teacher focus

Time: 20 minutes

1. **Support**: Do the activity on PB 3 p 13 with cube trains before children do the page itself and leave the board with ordered number bonds on display.

Working more independently

A. PB 3 p 13
B. '*Make cube trains of 10 and split them in several ways, eg 6 and 4. Keep all your splits and bring them to review time.*'
C. Use RS 9 to record different splits of 10.
Maths table:
D. **Extension**: '*How many balloons altogether on the IP? How will you make sure you don't count a balloon twice? Write the answer on the IP.*'

Whole class review

Time: 10 minutes

| Key idea | Pairs that make 10 are $9+1$, $8+2$ etc. |

- Ask questions to assess children's grasp of number bonds to 10. *'How many more do I need to add to 9 to make 10?' 'Tell me some pairs of numbers that make 10?'* Do MM 12 Five and ten whispers, again focusing on particular children for assessment.

- Ask questions to encourage children to think about the number pairs for 10 and to emphasise how you can reverse the order of numbers in addition, eg *'Tell me about adding the numbers 3 and 7, and 7 and 3.' 'Let's add 7 and 3 … that makes 10. What about the other way round?'* Establish that whatever the order, the answer is the same. (You can't reverse numbers for subtraction and get the same answer.)

Alternative starters

- Say rhymes about 10, eg *Seven Dizzy Dragons* p 14 'The big red bus', p 4 'Hippety hop' with 10 sticks of candy and p 23 'Chook, chook'.
- Play CG 32 Make up to 10.

Other lessons from this page

- Telling number stories
  Use the groups of creatures to the left of the teddies, eg there are 5 rabbits and 3 more come to play making 8 rabbits, or 8 ladybirds and 2 of them fly home leaving 6 ladybirds.
- Doubling
  double 3, 4 and 5 (turtles, snails, and frogs)
- Counting in 2s
  10 lots of 2 ants, 5 lots of 2 birds, 10 lots of 2 boots on the teds

IP 8 Playing by the pond

starts to link addition and subtraction to the number line/track.

Learning objectives:
- counting on and back along a number line and linking this to language of addition and subtraction
- counting, naming and ordering numbers 0–10/20

You need:

a large number track (at least to 10 with space to be extended) on the floor, number cards zero to 10/20, small number tracks such as RS 1, small objects, eg toys/cubes, dice, spinners, counters

Key words:

number names zero to 20 and beyond, count on/back, add, altogether makes, order, first, second, etc

Initial assessments:

Note who
- can count with one number name for each step (one-to-one correspondence).
- can recite the number names in order to 5/10/above.
- can take steps appropriately along a number line, not counting the space they are on.
- can recognise numbers, name them, order them to 5/10/20.

Lesson 1: Clown steps

Level: W–2

Objectives:
- counting on and back along a number line and linking this to language of addition and subtraction
- ordering numbers

Whole class starter

Time: 10–20 minutes

1. Gather the children around a large floor number track. Choose a confident child to walk along the track, counting out loud, one number word for each step.
2. Move children to where they can all see the number track on IP 8.
   Use a small toy and stand it on, say, 3. *'Look at the frog. What number is it standing on?'*
   'Suppose it hops along 2 spaces.' Hop slowly and deliberately along 2 spaces.
   'One hop, two hops. What number is the frog standing on now?'
   Give each child a small number track to 10 and a small toy. Ask children to repeat this on their number tracks.
3. Do several more examples on the IP. *'Start on 2 and hop on 2. We land on 4. 2 steps and 2 more steps; that's 4 altogether. Tell me what 2 bricks and 2 more bricks makes altogether.'*
4. Now, or in a later lesson, focus on the order of the numbers. Use numbers appropriate for your children (writing in 11–20 if needed). Recite the number names several times. Order number cards 0 to 5/10/15/20.

Group time

Teacher focus

Time: 20 minutes

1. Practise taking steps along the floor number track and extend it to 20.
2. Use the language of addition and subtraction: '*Start on 4, take 2 steps and we end on 6.*'

Working more independently

A. Let children take turns hopping along the number track on the IP, writing over the numbers. Let them practise writing in the teen numbers.
B. In pairs play PB 3 cover snake game.
C. '*Make up a game on your own number track.*' Have dice and spinners available.
D. **Maths table**: Have number cards 0–10, 11–20, sorting trays and toys. '*Count out 11 sheep, 12 dogs etc. Put the relevant number card by each group. Bring to review to be checked.*'

Whole class review

Time: about 10 minutes

Key idea	Numbers come in order on the number line.

- Recite number names to 5/10/15/20.
- Talk about the games played. Invented games can be taught to other children.
- Do a few minutes of mental maths. '*Shut your eyes and think about the number 4 on the number track. Which number will you be on if you count on/forwards 1 step?*' '*Which number is 1 smaller than 5?*' '*3 fish and 2 more fish. How many fish altogether?*'

Alternative starters

- Ask confident children to write in 11–20 on the track on IP 8.
- Talk about the race at the bottom of the page (ordinal number). (See 8e in the nursery section, p 56.)

Lesson 2: Mental maths on the number line

You need:
number cards 0–10

Level: 1–2
Objective: • early mental addition and subtraction

Key words:
number names 0–10, how many more to make…, add, plus, take away, equals

Whole class starter

Time: 10–15 minutes

1. Lay out the numbers 0–10 and draw attention to the zero. '*Close your eyes and pretend you are standing on number 2 on a number track. How many steps would you need to take to get to 4?*' '*What is three more steps on from 1?*'
2. Let your questions get more abstract. '*What is 4 count on 1? 3 steps back from 5?*' '*3 and 2 more?*' '*What is 2 plus 2?*'

Lesson 3: Ten and a bit more

Level: W–2
Objective: ● counting, naming and ordering numbers 0–10/20

Whole class starter

Time: 20 minutes

1. Recite numbers to 20. You, or a child, write in the numbers 11–20 on IP 8.
2. Give out a set of 11–20 number cards, one card to a child. Call these children out to the front and order them and recite 0–20. (Also have children with cards 0–10 if that is necessary.)
3. If children are not familiar with sets of 10, make some 10-rods with cubes, stick them with sticky tape and keep in a box labelled '10-rods'. Use these at mental maths time: *'One 10-rod and 2 more cubes, how many does that make?'* (12)
4. Seat the children in a circle. *'Hold up 10 fingers. 10 and 1 more makes 11. You don't have enough fingers! Join with the person sitting next to you and use 4 hands.'* Continue on to 20. *'11 and 1 more makes 12. Hold up 12 fingers,'* etc.
5. Emphasise names and how each number is written. Although in English we say 'seventeen' we write the 7 after the 1 (ten).
6. Muddle up cards 0–20 or 10–20 and sort them as a whole class. Recite the number names 0–20.
7. Show children PB 3 p 3. Discuss what to do: Count the geese (with long necks) in the picture, but not the goose next to the box where they write the number (12). There are: 11 horses, 13 goats, 15 cows, 19 flying birds, 20 ants, 18 flowers, 14 ducks (short necks), 17 sheep, 16 hens.

Group time

Teacher focus

Time: 15 minutes
Support: Help children with counting to 20. Use ten-rods and farm animals.

Working more independently

A. PB 3 p 3
B. **Maths table**: Make more animals for farmyard. Give a specific number, eg make 12 geese.

Whole class review

Time: 10 minutes

Key idea	13 is 10 plus 3, etc.

- Recite 0–20 and order number cards.
- Check that children understand that 13 is 10 and 3.
- End the lesson by playing CG 19.

Lesson 4: Number around us

Level: 1–2
Objective: ● writing numbers in real contexts

Whole class starter

Time: 20 minutes

1. Play CG 16 Washing lines.
2. Count all the class every day for a couple of weeks and record the numbers. Count other items to 20, eg children having sandwiches, the farm animals.
3. Talk about house numbers, bus numbers etc. Name the numbers. Write them up on display.
4. Match numbers around the class to counted items. *'Look at the 12 on the clock. Everyone count out 12 teddies/cubes. Put a matching number card by your cubes.'*
5. *'In pairs, muddle up then order a set of 0–20 number cards.'*
6. Explain PB 3 p 4. The bat, bear and engine have price labels to be completed, and children can record house numbers that they know. Read 'I am x years old.'

You need:

number cards 0–20, washing line and pegs, counting objects, counters, dice

Key words:

number names 0–20, first, before, next, between, in order, what comes next?

Group time

Teacher focus

Time: 15 minutes

1. Give support with teen numbers and do PB 3 p 4 together.

Working more independently

A. PB 3 p 4
B. Make a snake like that on the PB cover, but to 20, or above. Invent a game.

Whole class review

Time: 15 minutes

Key idea	Numbers can be used as labels.

● Talk about children's work from the teacher-independent activities.
● Name and write numbers children have used on the bus, doors, price labels.
● Recite numbers 0–20.
● Find numbers 0–20 on hundred squares IP 15.

Further lessons from this page

● Halving numbers: There are 4 butterflies (2 on flowers, 2 flying), 8 frogs (half in half out of the pond), and 10 mice (5 in a hole, 5 out).

IP 9 Robot world

gives early experience of numbers up to and beyond 20 to clarify place value and of 2D shape.

Learning objectives:

- numbers to 20 and counting on/back with the number line
- 2D shape
- language of position and movement

You need:

washable spirit pen, cards 0–20, washing line and pegs, large floor number track, numbers 0–20 on display (on the washing line or let children make a number frieze, see Group time A)

Key words:

number names zero/nought to twenty, count on/back, pattern, lots of (as quantity), how many altogether?

Initial assessments:

Note who

- can count to 20 by rote getting words in correct order.
- can count to 20 with one-to-one correspondence.
- can order numbers zero to 20.
- needs help with naming teen numbers.

Lesson 1: Roller coaster

Level: 1–2
Objective: • counting and ordering numbers to 20

Whole class starter

Time: 10–20 minutes

1. Write in the numbers 0 to 20 on the roller coaster spaces.
2. Play CG 16 Washing lines with numbers to 20. Recite numbers to 20 several times.
   'Now let's say the numbers very softly / every other number very loudly / clap loudly when we get to 20.' 'Now shut your eyes and say the numbers to 20.'
3. Play CG 3 Finger counting.
4. Use a small toy to count on and back along the roller coaster, eg *'Ten count on 2 is 12. Ten count on 3 is 13.'*
   'What number is one more / one less than 12?'
   'Which number is just before 11?' 'Which number comes between 18 and 20?' 'Name a number that is bigger than 10 and smaller than 20?' 'Let's count back from 20 and stop at 9.'
5. Introduce the maths table activity by numbering the front robots in order 1–7. *'What number robot would be next? Tell me a number that comes between 1 and 7?'*
6. Now or at a later lesson replace the roller coaster numbers with 10 to 31. Use this sequence to show the pattern of the numbers.

Group time

Teacher focus

Time: 15 minutes

1. Work with children on the large floor number track, stepping forwards and backwards.
2. Count out 20 objects with each child and relate the numbers to the roller coaster and the floor number track.

Working more independently

A. **Maths table**: Have ready cut 2D shapes and glue to make robots. '*Make a robot number track for the wall. Try to go beyond 30.*'
B. Use the IP to count some things, eg roller bladers (7 on the path plus 2 on the zig zag path). '*How many skates?*' (14 + 4 = 18) '*How many wheels on roller blades?*' (9 lots of 6, 54 altogether). There are 11 bugs, 5 with 5 sides and 6 with 6 sides. Each bug has 6 legs, 66 legs altogether. Write the numbers on IP 9.
C. **Extension**: Use a number track to 100 or a 100 square (IP 15) and circle all the numbers you can name.

Whole class review

Time: 10 minutes

Key idea	There is a pattern in the naming of numbers above 20: the unit digit repeats every 10.

- Ask small groups to count together to 20 so you can assess them. Hold up number cards and ask children to identify the numbers.
- If some children have counted up to and beyond 100, ask everyone to do that counting together. (Use IP 15.) Even children who are still struggling to get to 20 will benefit from hearing the pattern of the words up to and beyond 20 and can learn them by rote over the next few months.

Alternative starters

- Give each pair of children a shuffled set of cards 0–20 and ask them to sequence them.
- Move children on to stepping on or back along the number line, using the names of the larger numbers, eg '*Sophie stand on 13. How many steps to get to 14? What number comes before 14? What number comes just after 14?*'

Lesson 2: Shape robots

Level: 1–2
Objective: ● recognition and use of language of 2D shape

Whole class starter

Time: 10 minutes

1. Play CG 42 or 45.
2. Point to shapes in the top border, eg the circle and ask *'Come and point to another shape on the page that looks like this.'* *'Tell me about this shape.'* You are looking for the language of shape rather than naming shapes at this stage, but many children will know names such as circles, squares and triangles. Match shapes from the border to the picture, eg the house roof shape and the semi-circles. *'Can you find a big circle? What about a small one?'*
3. *'Let's count the sides on this triangle. Come and find another shape with 3 straight sides.'* *'Let's put a red dot on all the shapes we can find with four sides.'* (roofs, roller coaster carriages, windows, square and rectangular heads and bodies) *'Now let's put a blue dot on all the shapes with more than 4 sides.'* (bugs at the bottom)
4. *'Point to a robot with a round head. Can you find another robot with a round head?'*
   'Come and point to a robot that has a round body and a square head.'
   'Can you see two robots that are nearly the same?' (The robot flying the kite and the one sitting on the left in the green carriage both have round heads and square bodies.)
5. *'I'm thinking of a robot and it has a yellow triangle/3 sided body and a red head with 4 straight sides. Who can tell me which robot I'm thinking of?'*

Whole class review

Time: 10 minutes

> **Key idea** | Triangles have 3 straight sides etc.

- As children describe their work encourage others to ask them questions about it.
- Try to help children to generalise about shapes, eg *'Tell me what is the same about these shapes.'* (indicating some different shaped triangles)

Alternative starters

1. *Seven Dizzy Dragons* p 9 'Imaginary pictures', or p 29 'Square dance'.
2. CG 46 What's in the bag?, using 2D shapes.

You need:
2D shapes, pens

Key words:
flat shape, square, circle, triangle, rectangle, star, diamond/kite, side, corner, straight line, curved line, point, straight edge, curved edge

Group time

Teacher focus

Time: 20 minutes

1. Match plastic shapes to the small robots going up the mountain. (Alternate body shapes are circular and 4-sided, 8 round bodies altogether, 7 quadrilaterals.) Use this as an opportunity to describe the features of the shapes, eg the number of corners.

Working more independently

A. **Maths table**: NCM shape stencils, 2D shapes. Ask children to draw or cut out their own shape robots.
B. PB 1 p 10

Lesson 3: Where is it?

Level: W–1
Objective: ● the language of position

Whole class starter

Time: 10 minutes

1. Play CG 45 Where is it? Show children the word on a card if that is appropriate for your children.
2. Talk through with the children what they can see on the picture and note the words for position that children recognise, eg *'Show me the robots going down the zig zag road.'* Children can come out and mark items with the pen, or draw things.

You need:

model car or robot, a box, a toy roadway (a line of bricks will do), small toys/sorting items, construction toys, acetate and pen, words on cards

Key words:

over, under, above, below, on, in, outside, inside, in front, behind, next to, between, up, down, across, along, to, from, towards, away from

Initial assessments:

Note who

● can use simple words for position and respond appropriately.
● needs further help with some words.

Group time

Teacher focus

Time: 15–25 minutes
For children who need help with position words. Give tasks that illustrate words that children seem unclear about, eg *'Hide a teddy under the cot in the play house,'* or *'Stand monster behind the board.'*

Working more independently

A. **Maths table**: Polydrons or other small construction toys, margarine tubs, cars, teddies etc and trays to carry models to review time. You could give specific tasks such as *'Make a car going under a bridge.'*
B. Play hide and seek, one child shutting their eyes and another hiding toys. Give one clue, eg *'The red car is behind something blue.'* Towards the end of the session ask them secretly to hide something and give clues at review time.

Alternative starters

● Start a lesson with children in a large space where they can position themselves according to your instructions, eg sit on top of the mat, get behind the wall bars.

Whole class review

Time: 10 minutes

Key idea	There are words which describe where objects are.

● Find the hide and seek toys.
● Sort out the words for position into the ones most know and the ones they need to use more. Explain that you will be using these words in the next PE lesson.

You need:
cubes and other counting objects, number tracks

Key words:
how many altogether?, makes, put the larger number first, steps on a number line, in your head, picture in your mind, double, nearly double

Initial assessments:
Note who

- can explain what they are doing, and who still needs more help with what adding two numbers together means.
- shows some signs of more advanced stages of calculating (eg using known facts; see Mental maths, 'Teaching children the language of calculating' on p 14).

IP 10 How do you do it?

encourages children to talk about their mental strategies for counting and calculating.

Repeat these lessons several times to encourage children to talk about their mental strategies for counting and calculating. Read the Mental maths section 'Teaching children the language of calculating' on p 14 for further detail on how to teach this important lesson.

Learning objectives:

- exploring children's own mental strategies
- talking about and explaining new strategies

Lesson 1: How could you do it?

Level: W–1
Objective: • introducing the language of calculating strategies

Whole class starter

Time: 20 minutes

1. Take a simple calculation that could be done by the addition of near doubles, eg '*2 teddies and 3 more teddies would make how many teddies altogether?*'
   Draw the 2 teds and 3 more in the orange panel between the think clouds. Ask children to do it in their heads or with their fingers/cubes. Observe who is using which method.
   '*Tell me what you did and how you found your answer.*' Listen to and observe children. Give an analysis of what the child said and did, eg '*So, Lucy, you held up 2 fingers and then 3 more fingers and counted them, 1, 2, 3, 4, 5.*'
2. Draw Lucy's method in Ted's think cloud – perhaps drawing the fingers she held up.
3. '*Did anyone do anything different?*' (You may need to tell children what you saw someone do, to encourage them to join in.)
   '*Danny got 2 cubes and then 3 more and put them in a line and counted them starting from 1.*' Draw this in Dragon's think cloud.
4. '*I expect someone else did something different.*'
   '*Suzie said she knew 2 and 2 was 4, so 2 and 3 is one more than that.*'
   (At first it is likely that your 'different methods' will be different equipment, ie using a number track, cubes, fingers.)
5. '*Look at these different ways of doing that sum! You are all so good at maths.*'

Group time

Teacher focus

Time: 15 minutes

1. Repeat the starter activity with a small group to analyse strategies to find 3/4 'different ways', and record them in the four think clouds for review session.

2. In the small group you will be able to observe more closely so that you can identify children who are able to talk clearly about what they do in their head. These children can help you to move the rest of the class on. Many children will need time to see what to do when asked to talk about their 'in their head' pictures.

Working more independently

A. Ask children to draw what they did on plain paper. Provide interesting pens and cut-out think clouds in the four colours used on the IP page. Keep number tracks, cubes etc freely available.

B. **Maths table**: Play games in which a total score needs to be kept, for example toss bean bags into a box and score a point (or 2 points) for each bag that goes in. This gives practice in mental addition. Tally the score or mark it on the border number line.

C. **Extension**: Write numbers as far as you can go on the border number line on IP 14.

Whole class review

Time: 10 minutes

Key idea	Describing how to do calculations.

- Discuss children's drawings with them, using the appropriate language.
  'Tell us what you did here, Tom. Did you count them one at a time?'
- Do a few 'in your head' calculations, eg *'Shut your eyes and think of 2 teds. Now 2 more come to play. How many altogether?'*

Alternative starters

- Read 'Five little astronauts' to show one way to talk about adding 1, and/or 'Ten little teddies' as an example of counting back in 1s (*Seven Dizzy Dragons*, pp 3, 8).
- Take another simple calculation such as 4 take away 2, and repeat the activity.
- Towards the end of the year do CG 27 (I can double numbers) and move children on to learning to add using near doubles, eg *'If 3 and 3 is 6, then 3 and 4 must be 7.'*
- Later start by using standard notation rather than pictures in the space between the clouds, eg 2 + 3 =

Lesson 2: Do it both ways

Level: 1–2

Objectives:
- explaining counting back and counting on
- talking about more than one way to do a calculation

Whole class starter

Time: 15–20 minutes

1. Practise using a large floor number track. You or a child demonstrate a 'sum' on the number track, eg 7 – 4. (The words 'the difference between' are appropriate here and this can be illustrated by ages of children, eg *'You are 4 and your brother is 7. What is the difference in your ages?'*) *'Stand on 7, count back to 4. How many steps?'* *'Now stand on 4 and count forwards to 7. How many steps?'*

2. Use two of the think clouds on the IP to draw stepping back and stepping forward.
   '7 step back to 4, land on 6, 5, 4. Three steps.' *'Start on 4 and step forward to 7, land on 5, 6, 7. Three steps. Three steps each time.'*

3. Explain that counting on from 4 and counting back from 7 gives the same answer.
   'We did the same sum in two different ways.'

Group time

Working more independently

Time: 10–15 minutes

It would be appropriate to follow this starter by asking each child to make an individual drawing of what they did in their head for one specific calculation, eg 10 take away 6. Ideally, use plain paper as this can help children to explore their own mental images. Blank number tracks, cubes etc need to be freely available. Unifix number lines can be used by children with poor recording skills.

You need:

a large floor number track, things to count, number cards 0–10

Key words:

count on, count back, how many more?, difference, take away, add on, different ways

Initial assessments:

Note who

- can talk about their own method of calculating using appropriate language.
- can listen to what another child did and then repeat what was said.

Whole class review

Time: 10–15 minutes

| Key idea | There is more than one way to do a calculation. |

- Ask children to work in pairs (not a child they sat next to during the drawing time) and talk about and compare what they did with their partners. You need to observe and listen very carefully.
- Draw out from the talking some of the main points, eg children that show steps on number tracks, those that counted back and those that counted on, those that used fingers or cubes etc.
- Refer back to IP 10 and emphasise that there is often more than one way of doing a sum.

Other lessons from this IP

1. Use the row of 5 boxes at the bottom of the page to write in, eg 3 + 2 = 5. There is also space for children or you to record or draw.
2. Use the border number line for counting in 10s and 2s, and to show how to record hops on a number line.
3. Explode a number. Put a number, eg 5, in the orange space in the middle. Ask children for different ways to make 5 (1 + 4, 6 – 1, etc.) and write these in the think clouds. Also ask children where they can see '5' in the classroom, eg on the clock, on the 5 gram weights.

Alternative starters

1. Repeat the lesson with different calculations. Ideally, this activity should be repeated a few times a term. It does not have to be a whole lesson; a few minutes of mental maths will provide good practice.
2. Play CG 35 Mice in a hole. Ask children to explain how they worked out how many were in the hole.

IP 11 Inside outside

practises splitting numbers and making number sentences.

Learning objectives:

- counting
- language of number
- number bonds
- using standard addition and subtraction symbols

You need:

small plastic teds/toys, quoits or paper with a circle drawn on

Key words:

and, add, together make, plus, take away, number names 0–10

Initial assessments:

Note who can

- count accurately.
- use a wide range of language of number.
- talk about number bonds/ splitting numbers.

Lesson 1: Telling number stories

Level: W–2

Objectives: • counting
        • language of number and number bonds

Whole class starter

Time: 15 minutes

1. Play CG 28 Inside outside.
2. Give each child 4 or 5 toys. The teds are playing a game hiding in and out of their house. '*Put 3 teds inside. How many outside?*' '*How else could you arrange the 4/5 teds?*' (eg 1 in the circle and 4 outside), '*If we add together all the teds inside and all the teds outside, how many teds will we have altogether?*' '*What if you have no teds inside. How many outside?*'

   Ask children to split up their groups of toys practically in different ways using a quoit/paper circle.

Group time

Teacher focus

Time: 15 minutes

1. Split a set of objects into two groups, eg 7 toys into 5 and 2. Use a wide variety of language of number to extend children's thinking and use of language, eg
'We all have 6 cubes. How could we split them into two groups?' Provide circles etc or set this in a context, eg buns on plates.

2. *'Now if we put our two groups back together again, how many buns will we have?' 'Now let's make some number sentences.'* (You could write these down to show at review time.)
'3 buns and 3 more buns makes 6 buns altogether.'

3. *'Start with 6, split them into 3 inside and 3 outside, but you still have 6.' 'Start with 6 and hide 3, then you have 3 left.' '6 can also be split into 5 and 1 or 4 and 2.' 'Are there any other ways to split 6?'*

Whole class review

Time: 10 minutes

| Key idea | Using a variety of words for addition, eg add, plus. |

- Whilst children talk about their work, encourage them to use the language of number sentences. (It is not important at this stage to focus on finding every possible combination of numbers.)

- Conclude with a mental maths session, again focusing on the language of maths. When they are ready, gradually move children from the more concrete *'3 inside, and 3 outside, how many altogether?'* to more abstract, *'What is 3 add 3 more?'* and *'How many do you have if you have 4 and 1 more?'*

Working more independently

A. PB 2 p 6

B. **Support**: Let children play a 'teds in and out' game with just 3 or 4 teds and a paper circle, doll's house, etc. *'Draw what you find out for review time.'*

C. Let children draw on the blank space on IP 11. Give numbers to match experience or let them choose: *'Draw at least 3 different patterns/arrangements of your number.'*

D. Let confident children show other ways to split up 4 and choose another number to split up on the blank space on IP 11.

E. **Maths table**: Small world play characters with either quoits, a pretend sand pit, paddling pool or teds house. Let children explore in and out numbers. Give children number cards to focus them onto the number of toys to use. You could provide circles of blue paper for the pool and other paper to stick or draw their work.

Alternative starters

1. Read 'Ten little teddies' *Seven Dizzy Dragons*, p8. Start with just 5 or 6 teds. Then let children tell stories that split the teds into two groups, eg 4 teds went out to sea, 2 went in for a swim, so 2 were left in the boat.

2. Do something similar with 'Seven dizzy dragons' p7. *'7 dragons, 3 run away, how many are left?'*

3. Repeat the whole class starter but draw 6 or 7 teds on the IP.

4. Repeat this important activity several times in different contexts, eg buns (*'2 buns on the red plate, 2 on the blue one'*) or the play house (*'3 children are inside the play house and 2 more are outside knocking on the door, how many children altogether?'*).

5. When children's language has developed, do the activity with 4 teds so they can then record on PB 2 p 6. *'What if we start with just 4 teds?'* Use the top right-hand space on IP 11 to demonstrate recording. Draw stick people or simple teds.

Lesson 2: Lily pads

Level: 1–2
Objective: • language of number and early addition (and subtraction)

Whole class starter

Time: 15 minutes

1. Play CG 23 Lily pads.
   Go around the circle varying the language for each addition. *'2 plus 1 is 3,' '4 add 2 makes 6,' '5 and 1 more equals 6 altogether.'*
2. Play in pairs. Each child puts frogs onto a lily pad, then both hop all their frogs onto a third lily pad. Go around the circle with each pair hopping their frogs onto the third lily pad. *'Jack has 2 frogs, Ella has 2 more. They all hop onto the last lily pad. How many altogether, Ella?'* Repeat this several times so that children understand that the combined number is on the last lily pad.
3. Introduce number cards for each lily pad. Each pair of children makes a new sum with numbers and tells all the class. Demonstrate the appropriate mathematical language by repeating and rephrasing what the children say.
4. Draw lily pads on the space under the teds on IP 11 just like on PB 2 p 12. Demonstrate recording. Give children who have problems with recording plain paper and ask them to draw what they have done.

<table>
<tr><td>

You need:

paper lily pads and frogs (cubes), several sets of number cards 0-10

</td></tr>
<tr><td>

Key words:

and, together makes, add, plus, equals, how many, more/fewer/less than, how can you check, same/different number

</td></tr>
</table>

Initial assessments:

Note who

- can use language of number and early addition.
- can add 2 numbers with cubes/fingers.
- will need more help with language and recording.

Group time

Teacher focus	**Working more independently**
Time: 15 minutes 1. **Support**: Play Lily pads again or change the context to, eg, ducks in a pond. Emphasise the appropriate language. Check children understand the combining of two groups. 2. Set in real life contexts. *'If you have 2 carrots for the rabbits and Jake gives you one more, how many altogether?'* Give children plenty of experience with a wide range of language.	**A.** PB 2 p 12 **B.** Continue playing Lily pads. Give each child 3 lily pads and two dice/spinners labelled with numbers to suit their ability. *'Find you own way to record what you did.'* **C.** **Maths table**: Play tiddlywinks, trying to flip green counters (frogs) onto a lily pad with a large counter. The winner is the first to get 2 frogs onto their lily pad.

Whole class review

Time: 10 minutes

Key idea	Combining two groups to make a bigger one.

- *'How many do you get if you have 4 and 1 more?' 'When Fran put her 3 cows with Annette's 4 cows, how many did they have altogether? Can you check that somehow?'*

Lesson 3: Recording addition

Level: 1–2

Objectives:
- early language of addition
- using standard addition symbols

You need:

small toys for each child

Time: 25 minutes

1. Give children cubes/toys to represent the birds, butterflies, balloons etc on the IP. *'There are 2 birds asking 3 more birds to come and play with them.'* Let them use cubes/fingers to show the numbers. You can inject the necessary language: *'2 birds and 3 more birds equals 5 birds altogether.'*
   Ask them to make up more number stories about the birds.
2. Show how to do that addition on a number line. Draw a line in the space on the IP or let children use small number lines.
3. Demonstrate the standard way of writing addition. You can do this on the IP, eg with the birds write '2 + 3 = 5'. Let children lay out sums with cards as in the lower part of PB 2 p 14. Introduce PB 2 pp 14 and 15.

Whole class review

Time: 10 minutes

Key idea	Number stories can be written in a quick way using standard notation.

- *'Tell me your number story and help me to write it in a quick way.'*

Alternative starters

- Another day, repeat the lesson, recording addition sums on the space on the IP. Make sure children read the sums using a wide variety of language, eg 3 + 2: *'If you have 3 cubes and 2 more cubes that makes 5 when you count them up altogether.'*

Lesson 4: Recording subtraction

Level: 1–2

Objectives:
- language of number particularly subtraction
- subtraction by crossing out
- using standard addition and subtraction symbols

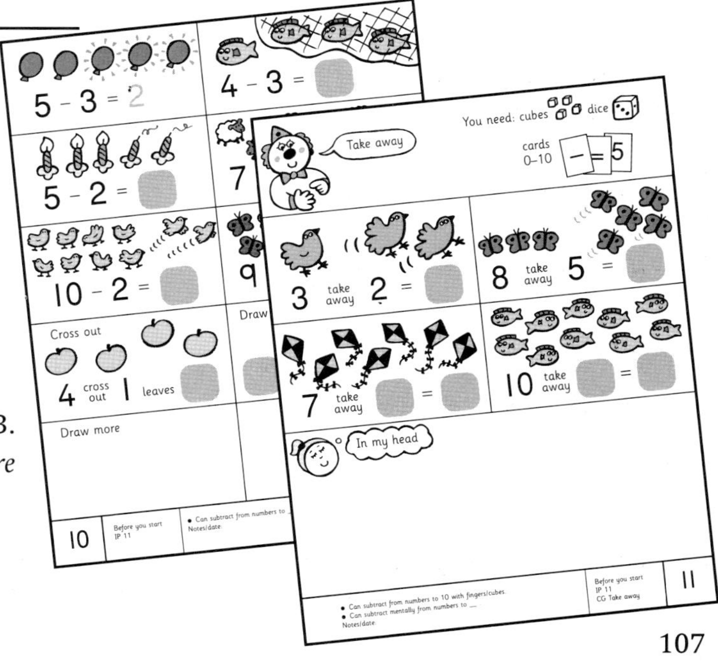

Time: 45 minutes

1. Start with a circle game for subtraction eg CG 33.
2. Use IP 11 to make up subtraction stories, eg *'There were 5 sheep then 2 ran away. How many were left?'*
3. Demonstrate crossing out for subtraction.
4. Introduce PB 3 pp 10, and 11 the next day.

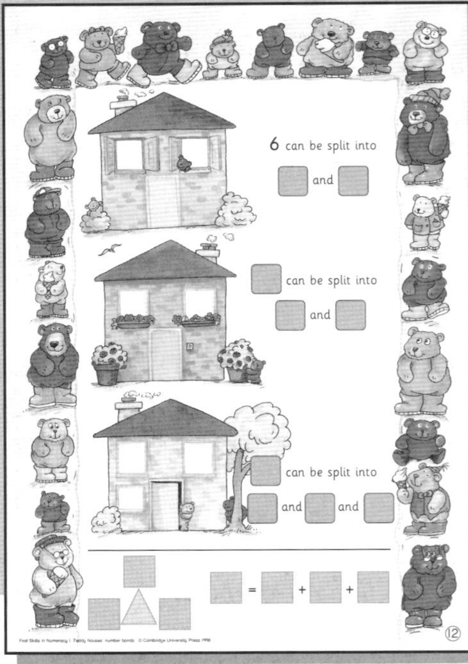

IP 12 Teddy houses

practises addition and subtraction, and focuses on finding unknown numbers.

Learning objectives:

- splitting numbers into 2 and 3 parts
- finding unknown numbers

Lesson 1: Teddy's front door game

Level: 1–2

Objectives: ● splitting numbers into 2
● finding unknown numbers

Whole class starter

Time: 10 minutes

1. Play CG 31 Teddy's front door game.
2. Show teddy's front door game on IP 12. *'Ted likes to play number games with his front door number, 6. 6 can be split into 2 and 4. So he puts those numbers on the windows. How else could we split 6?'*
3. Practise with other numbers (5 is used on PB 3 p 15) using paper ted houses, splitting into just 2 groups in the first lesson.
4. Demonstrate how to use the paper ted houses with number cards for the maths table.

You need:

cubes / acorns / teddies, paper ted houses (RS 8), number cards

Key words:

number names 0–10, add, together makes, plus, how many?, equals

Initial assessments:

Note who

- counts from 1 each time.
- can 'see' numbers out of sight.
- needs help splitting smaller numbers and more help with counting.

Group time

Teacher focus

Time: 20 minutes

1. Assess children's understandings of number bonds and set targets for future work. Use RS 8 and IP 12 to do more unknown numbers, eg *'10 on the door. 1 on this window. Which number goes on the other window?'*

2. **Support**: Ask children to count items from the teddy border on the IP. (See Nursery lessons for IP 12.) Check to see how their counting is progressing and try to find time to do more splitting of maybe 3 or 4 during the week. See CG 29 Splitting numbers.

Working more independently

A. PB 3 p 15

B. Children can explore larger numbers and splitting into 3 groups using RS 8.

C. **Maths table**: A4 size paper/card houses with a door and 2 windows, Blu-Tack and sets of number cards. *'Play teddy's front door game and bring your houses to review time.'*

D. Make necklaces with 2 colours of beads or 'trains' with 2 colours of cubes using 6 beads or cubes each time.

Whole class review

Time: 10 minutes

Key idea	You can find unknown numbers by counting on or counting back.

- *'If there is 10 on the door and 3 on one window, what is on the other window? How did you work it out?'*

- *'What did you learn today about numbers?'* Expect children to begin to know some number bonds, eg 4 and 2 is 6, 3 and 3 is 6, by heart. Emphasise that they can use what they know for adding other numbers, eg 2 and 3 is double 2 and 1 more.

Alternative starters

1. Start with another number. Knowledge of number bonds of 10 is very important so do the activity with 10 several times towards the end of the reception year.

2. Start with a mental maths game: *'I'm thinking of a number in my head. It can be split into 6 and 1. What's my number?'*, or MM 12 and 13.

- Relate this work to the number track. *'Stand on 6. Take 2 big steps to get to zero,'* eg one step 4 to 2, then 2 more to zero.

Lesson 2: Ten can be split into …

Level: 1–2

Objective: ● splitting numbers into 3 groups (and reading an equation appropriately)

Whole class starter

Time: 10 minutes

1. This activity uses the third house, with three windows, on the IP, and the lower border. Start with practical splitting of cubes as in lesson 1 but this time split 10 cubes into 3 groups. Ask children to tell you some sentences about their groups, eg '*I split 10 into 1 and 3 and 6, and if you put all of them back together again you still have 10.*'
2. Give children cubes and paper ted houses. '*Find more ways to split 10.*' Vary the language. '*I have 6 in one group and 2 in another. How many in my last group? 6 and 2 and how many more make 10?*'
3. Record this on the house to remind children of earlier work.
4. Show how to use the triangle in the lower border for recording. Put 10 in the triangle and record one child's 3 groups in the corner spaces.

You need:

cubes / acorns / teddies, paper ted houses with 3 windows (RS 8 amended)

Key words:

number names 0–10, add, together makes, plus, how many?, equals

Initial assessments:

Note who can

● split numbers and accurately count groups.
● use a wide range of language of number.

Group time

Teacher focus

Time: 10 minutes

1. Record on the equation to the right of the triangle, eg '*10 can be split into 1 + 3 + 6*'. Use a variety of appropriate words to explain the = sign, 'balances', 'equals', or 'can be split into'. (We are not suggesting at this stage that children record this themselves, but we are building up confidence with standard symbols used in equations and removing the idea that all 'sums' are in the pattern 3 + 4 = 7. This will be developed in later years.)

Working more independently

A. Choose numbers and split them into 3 groups. Record on RS 8.

B. **Extension**: '*Working in pairs, take 3 number cards from a box and add them up, making cube 'trains' in 3 colours. Bring the 'trains' to review time.*'

C. **Support**: (adding small numbers) Give children more practice with adding numbers 3–6 by doing activity B above, adding just 2 numbers. They can 'record' this with cube 'trains' in 2 colours.

D. **Maths table**: Children choose coloured paper triangles that either have numbers in the middle to be split up or are blank for children to fill in from number cards. They then split these numbers of objects into 3 groups and write how many there are in each of the corners of the triangle.

E. Paper ted houses with 3 windows, number cards etc, for children to repeat the activity in a different context.

Whole class review

Time: 15 minutes

Key idea	Numbers can be split into 3 as well as into 2 groups.

- Go over the language needed for this task.
  'Tell us about one of your numbers and how you split it into 3 groups.'

Other lessons from this page

- **Looking for pattern**
  Use the border and chant the colour pattern of the bears aloud: 'Blue, red, blue, red, ...' Say the size pattern (it is easier to start on the first little bear) 'Little, middle, big, little, middle, big...'
- **Solve simple number problems**
  Use the border to develop simple problems. *'Put a blue cube on all the teds with ice cream* (6 teds) *and a red cube on all the teds with hats* (5). *Do more have ice cream or do more have hats?' 'Do any teds have hats and ice cream?'* (Yes, 2.)
- **Doubling (and halving) numbers (extension)**
  Use the two houses with two windows and put even numbers on the door, 'Double 3 is 6.' The ability to double and halve is important for Y1.
- **Counting in 2s**
  Count the wellington boots. (22 bears, 44 boots)

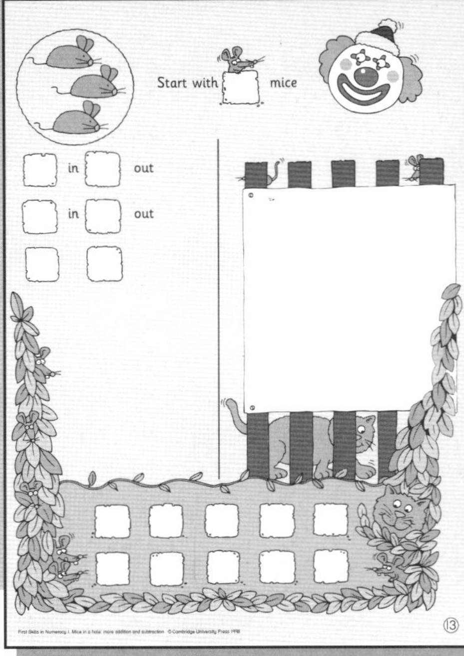

IP 13 Mice in the hole

develops more abstract language for addition and subtraction.

Learning objectives:

- splitting numbers
- mental addition and subtraction
- number bonds to 10
- experience with language and symbols of early addition

Lesson 1: 4 mice in a hole

Level: 1–2
Objective: ● splitting 4 and early addition and subtraction

Whole class starter

Time: 10 minutes

1. Play CG 35 Mice in a hole.
2. '*Close your eyes and see if you can see a picture of the mice in their hole. There are 4 mice altogether and 1 of them is outside the hole. How many are still in the hole?*' Ask children how they worked it out. '*We have 4 mice altogether and you can only see 1, so how did you know how many are left in their hole?*' '*Can you use your fingers to show us how you worked that out?*'
   As the language develops make questions a little more abstract.
   '*3 and how many more makes 4?*' '*What is 3 plus 1?*'
3. Send children off in pairs with their 'mice'. Encourage them to talk to their partner about what they are doing.
4. Children's recording can give us insight into their thinking if we let them record in their own way, so be careful as you record on the IP, eg 2 in and 2 out. '*On the PB page you can record in any way you want. Try to draw what you do.*'

You need:

cubes/plastic toys to represent mice, cloth

Key words:

and, add, makes, altogether, how many more to make?, plus, equals

Initial assessments:

Note who

- seems unable to calculate mentally with 4. (See support activity below.)
- with a little more support, might be able to do the task. (These children can make up the teacher focus group.)
- can calculate correctly.

Group time

It is appropriate for all the children to be doing variations of the same activity after the starter.

Teacher focus

Time: 20 minutes

1. Work with the children on PB 1 p 14 using cubes.

   Work one to one to make assessments of children's grasp of number concepts and the language of number. A suitable activity for this would be to repeat Mice in a hole or use a different context, eg kittens in a basket or sandwiches in a lunch box. *'How many kittens are still hiding? How do you know that?'*

2. **Support** (counting to 3 and splitting 3). Play CG 28 Inside outside with children who are still struggling with numbers to 3.

Working more independently

A. PB 1 p 14

B. **Maths table**: Set up a mouse hole and have a pot of 'mice' so that children can choose their number to work with. You could have a 'cat' that is on the look out for any stray mice and chases them back into their hole. Record on RS 4. (Mice can be made from individual sections from an egg box and a piece of wool for a tail.)

Whole class review

Time: 10 minutes

Key idea	Understanding the language of '… and … more'.

- Ask children to talk about their drawings. Model a variety of mathematical language, eg *'2 mice and 2 more mice makes 4 altogether.'* *'We have 4 mice, so 3 mice out of the hole leaves just 1 still in.'*

- Gradually move on to more abstract mathematical words as your children are able to cope with that, eg *'What is the total of 2 mice and 2 more mice?'* *'What is 3 and 1 more?'*

- Some may well have made some attempt at being systematic and finding all the different ways of recording the 4 mice, and others may try to do that. *'Look, Tamsin made a pattern, 4 in, none out; 3 and 1; 2 and 2; 1 and 3.'* It is not important if they can't see the number pattern at this stage.

Alternative starter

- Do the activity in a different context, eg eggs in a basket. *'There are 8 eggs in the basket altogether. If I bring out 1, how many are left in the basket?'*

Lesson 2: Cubes and mice

Level: W–2
Objective: ● splitting 5 and 6 and early addition and subtraction

Whole class starter

Time: 10 minutes

1. This lesson repeats the important content of lesson 1. Start with CG 35 Mice in a hole but this time use a cube 'train' of 5 cubes. Children record in their own way on PB 2 p 7. If children are not grasping the task, work with them to make different 5-cube trains, all made up of 2 colours, eg 1 red, 4 blues etc.
2. Repeat the activity with the IP and 6 mice. Children record on PB 2 p 8. Compare recordings at review time.
3. **Maths table**: Set the activity in other contexts, eg rabbits in a burrow, bears in and out of their house, and provide toys and cloth or box to hide them.

Lesson 3: Making sums

Level: 1–2
Objective: ● experience with language and symbols of early addition

1. Play CG 24 Train ride.
2. When children are using the language of addition reasonably well, use IP 13 to show how a sum is written, using numbers they have used for the circle game. Write either on the board or in the spaces at the bottom of the page, eg 4 + 1 = 5.
3. Use clown's board for 'words we know for adding', e.g. 'and', 'altogether' etc.

Working more independently

Children can do PB 2 p 13.

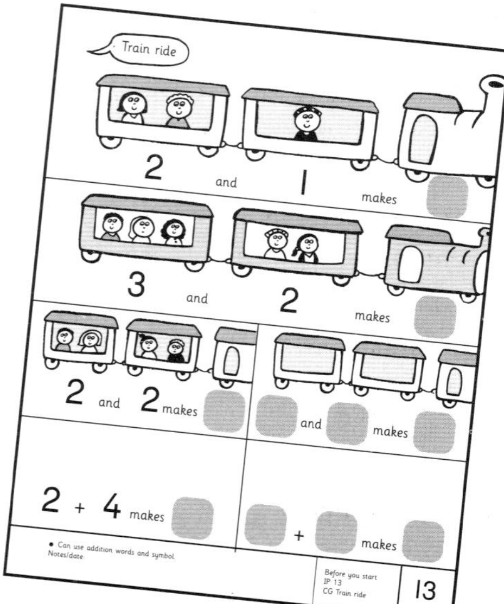

Lesson 4: Adding in your head

Level: 1–2
Objective: ● mental addition

Expect to repeat this lesson every day for a week or more.

Have a number sentence or sum for the day and ask all the children to read it to you two or three times in the day.

1. Play CG 22 Adding with cards and teddies, and give children plenty of experience with reading sums. Use the board or the spaces at the bottom of the page to write simple sums that children will be doing later in the PB, eg 3 + 2 =. Ask children to read that sentence to you and to work it out with fingers.

2. Play CG 26 In your head. Give children repeated practice with mental number using just small numbers until they are very confident.

3. PB 3 p 6 and p 7.

You need:

number cards 1–5, teddies/cubes

Key words:

and, add, plus, equals, total, together makes, more

Lesson 5: 10 mice escape from the cat

Level: 1–2
Objectives: ● number bonds to 10

1. Repeat lesson 1 with 10 mice. By now children should have a wide vocabulary of number. They should be starting to know the number bonds of 10 by heart.

2. Play CG 29 Splitting numbers with 10 mice.

You need:

cubes, cloth

Key words:

and, add, plus, makes, equals, together make, altogether, how many more to make …?

Working more independently

A. PB 3 p 12 again lets children record in their own way. Some by now can be encouraged to record things 'in order', especially if you have done finger counting in a pattern, eg none up, 10 down; 1 up, 9 down etc.

B. PB 3 p 14

Alternative starter

● Do MM 12 (10 whispers).

IP 14 Number lines and beads

splits numbers into 5 and a bit to demonstrate that, to add, it is not always necessary to count from 1.

Learning objectives:

- counting to 3/5/above
- visualising numbers as 5 and a bit more
- counting on and back on the number line
- language of addition and subtraction
- counting in 5s/10s

Lesson 1: Five and …

Level: 1–2

Objectives:
- counting to 3/5/above
- visualising numbers as 5 and a bit more

Whole class starter

Time: 15 minutes

1. Put the laced 1–5 beads/cubes in the feely bag and pass it around the group. Children take turns to take out one lace, hold it up and count the beads.
   'Now hold up that many fingers. Match one bead to each finger.'
   Try to get children so familiar with the numbers that they can gradually recognise them. *'Do you need to count them?'*
2. Show children how to record this on the IP.
3. Introduce numbers above 5. Pass the bag of laced beads 5–10 around the circle. Again match number to fingers. (See CG 5 Beads.)
4. Keep emphasising that the 5 red beads are always there, so they do not need to be counted each time. (Some children will take a while to grasp that.) *'Just count the blue beads. 5 red beads and 3 blue beads. How many is 5 beads and 3 more beads?'*
5. *'Shut your eyes and make a picture of the 6/7/8 beads.'*
6. Move on to 11, making the 11th bead red like the first 5 (or a third colour). Keep doing the circle games to give children repeated experience, using higher numbers when they are ready.

You need:

laces (all the same colour), red and blue beads/cubes, some ready laced beads or cubes as on the IP (1–5 at first, then adding the second colour to show 6–10), a feely bag, red and blue pens (plus later a third colour), 'trains' of cubes in those 2 (later 3) colours

Key words:

number names 0–10, and … more, add on, together make, same/different number

Initial assessments:

Note who

- can count to 3/5/beyond.
- doesn't need to count the 5 (conservation of number).

Group time

Teacher focus

Time: 15 minutes

Repeat the starter and work with children as they record what they do. Many children will need help to see that 5 red beads are still 5, however many blue beads you put after them.

Working more independently

A. Make a long line with beads or cubes with alternately 5 red and then 5 blue cubes. Make it as long as you want. (Unifix number lines are good for this.) '*Can you tell us how long your line is?*'

B. Make a long line of Polydron shapes, alternately 5 of one colour then 5 of another.

C. HB p 14

D. **Maths table**: Laces, beads in 2 colours (or cubes), number cards 1–10 (or 20/above to suit your children) with holes to be threaded onto the laces if you want.

Make necklaces / cube 'trains' with beads in 2 colours. String these up on display in order and use them at mental maths time. If the numbers are detachable it can make a teacher-independent interactive display.

Whole class review

Time: 10 minutes

> **Key idea** | Numbers can be split into 5 and something.

- '*Show me 7 on your fingers. So, 5 and how many more do you need to make 7?*'
  '*5 and how many do you need to make 8? Show me that with your fingers.*'
- Repeat CG 5 asking a child to lead it.
- Use fingers to go from 1 to 10 or above, as on the IP.

NB to show the pattern, on the IP fingers are shown turned down in order, but it is very hard to do numbers such as 9 as shown. Talk about this.

Alternative starter

- You will need to repeat this lesson several times at different levels, starting each time with CG 5 Beads. Follow up with recording: PB 1 p 3; PB 2 p 5; PB 3 p 2.

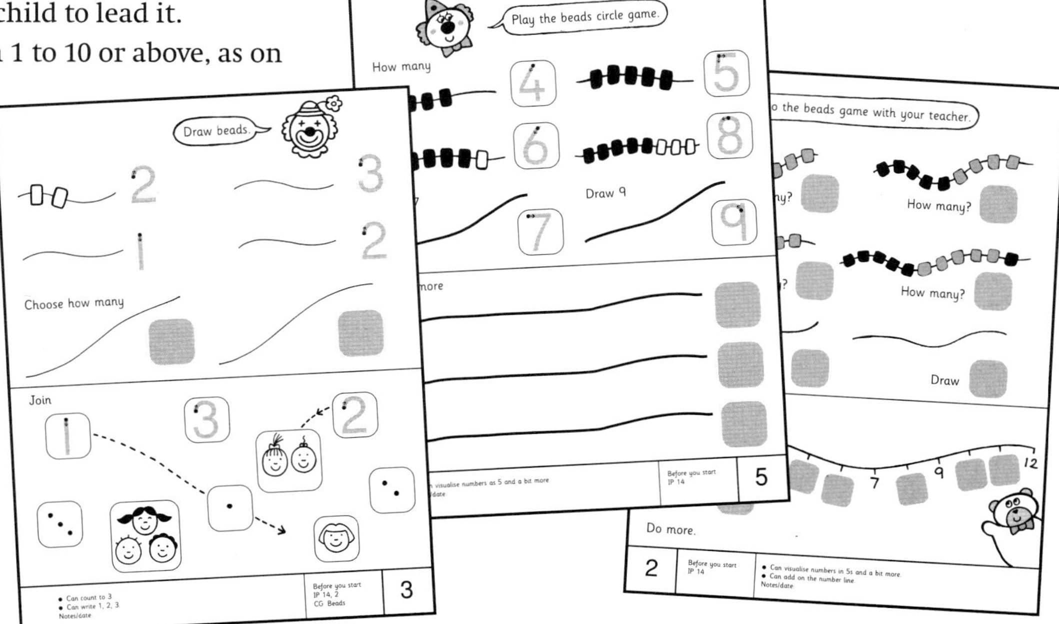

Lesson 2: Counting along the number line

Level: 1–2

Objectives:
- counting on (and back) on the number line
- ordering numbers
- language of addition (and subtraction)

Whole class starter

Time: 15 minutes

1. Using a set of number cards 0–10, ask children to set out the cards in order, on the carpet. Recite 0–10 if children need that.
2. Demonstrate steps on the IP, using the number track or bridging number line.
3. Play Step on, step back: Stand a teddy on the number track, then make it take 1 or 2 steps forward. *'Look at ted. She started on zero and stepped on 2 spaces. Where did she end?'* *'1 step from 2, end on …?'* *'2, step on 2, where will you end?'* *'If teddy starts on 2 and steps forward 3 where will she end? (what number will she end on?)'*
4. *'Shut your eyes and see a picture of a number track in your head. Choose a favourite toy and stand it on 2. Now make it take 1 step. Tell us the number your toy is standing on.'* Make the link between '2 and 1 more step' to '2 plus 1 more object'. *'Start on 2, hop on 3, land on 5. Now let's do that with our fingers/teddies. What is 2 teddies/lollipops and 3 more?'* On the lower part of the IP record sums to match with hops.
5. Repeat for subtraction and counting back.
6. Practise plenty of mental calculation. *'Anna is 4 and her big brother is 8. What is the difference in their ages?'* (Count up from 4 and back from 8.) *'Mum made 6 buns. 2 were eaten, so how many buns are left?'* (Count on and back on the number track.)

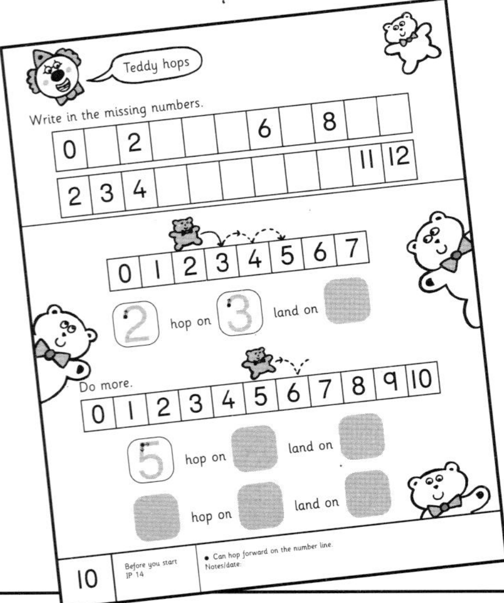

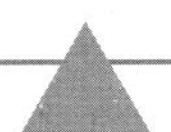

Group time

Teacher focus

Time: 15 minutes

1. **Support:** Do more work with children taking steps along small number tracks and help them to use the language of addition/subtraction, then record their work on PB 2 p 10.

Working more independently

A. PB 2 p 10

B. **Maths table:** Put out small number tracks and teds. Children can practise taking steps then record on the number tracks.

Whole class review

Time: 10 minutes

| **Key idea** | Addition can be done by counting on and subtraction by counting on and back. |

- Take steps on the floor number line using a variety of language, eg *'Count on/forward/up towards 10.'* It is important to repeat this mental maths work on the number line every day for a few days.
- *'What do you know about numbers now? What would you like to learn next?'* This can give insight into children's developing thinking.

Alternative starters

1. Step in 1s and 2s along the number track, eg *'Start on 3, take 1 step, where will you be?'* Introduce the language of addition after doing the activity a few times, eg *'If you take 1 step on from 4, you land on 5'*. *'What number is 1 more than 4?'* *'4 beads and 1 more bead is how many altogether?'*

2. Step forwards and back in the same lesson to show how stepping back undoes what stepping forwards does. *'Start on 4 and step forward/on 1. Where are you now? (5) Now start on 5 and step back 1.'* Repeat this with different numbers and then ask *'What is happening?'* (You get back to where you started.)

   Children can do PB 2 p 11.

Lesson 3: High fives!

Level: 2
Objective: ● counting in 5s/10s
Time: 15 minutes

1. Use the lower border to count in 10s. Recite the numbers and link with the hundred square on IP15. Read 'The chimpanzees' tea party' *Seven Dizzy Dragons*, p12.

2. Demonstrate how to count in 5s with lots of hands or the hands in the lower border.

3. Recite the numbers 5, 10, 15, 20 etc with the children. See 'Number patterns' *Seven Dizzy Dragons*, p24.

Working more independently

A. **Maths table**: Provide number tracks, 10-strips (or other ten-rods), IP pens, and encourage children to explore the border.

B. Draw around your hand (or do hand printing). Help with cutting out. *'Let's make a counting in 5s/10s hand line'* (as at the bottom of the IP). Put out number cards 10, 20, 30 etc so children can create an interactive display.

IP 15 Washing lines and hundred square

uses the hundred square to develop place value and number patterns.

Learning objectives:

- counting to 10/100
- recognising and ordering numbers to 10/20/100
- adding and subtracting 1 to/from a number
- counting in 10s and 2s

Lesson 1: Experiencing larger numbers

You need:

cubes, ten strips/trains (lines of 10 cubes held together with sticky tape), washing line, clothes pegs, number cards 0–10/20/100

Key words:

number names 0–20, before, after, close to, larger/smaller, order, next, between

Initial assessments:

Note who

- can count above 10.
- can recognise and point to familiar numbers beyond 10.
- can order numbers to 10/20/100.

Level: W–2

Objectives:
- counting, recognition and ordering numbers to 10/20/100

Whole class starter

Time: 15 minutes

1. Do MM 6 The hundred square to make sure all children understand how a hundred square works and show how it links to a number line.
2. Play CG 16 Washing lines and circle some numbers (eg house numbers, favourite numbers, numbers of children in each class in the school) on the 100 square.
3. Count out cubes to show some of the numbers, eg 21. Use 'ten trains' to demonstrate numbers. Show 100 and let children talk about that. (lots, more than I can count, etc)
4. Use a wide range of language. *'Tell me a big number close to / almost 100.' 'Does 3 come before or after 5?' 'Which is closer to 100, 3 or 89?' 'Is 30 larger or smaller than 50?'* Indicate the positions of these numbers and pick out any near them already identified by the children, eg *'22 was Ben's favourite number. Is it closer to 3 or 89?'*
5. Order numbers to 10 using number cards on the washing line.
6. Shuffle number cards to 10/20. Deal out five cards. Order the numbers and write them on a washing line on the IP.

Group time

Teacher focus

Time: 15 minutes

1. Help children to order beyond 10 and 20 using the 100 square and number cards.
2. Extend the class number line at least to 20 and help children to find numbers they know on it.

Working more independently

A. PB 2 p 9

B. **Maths table**: Provide paper circles to make paper caterpillars as on PB 2 page 9. Let children extend these as far as they can go. They can be displayed and used for several days for counting at mental maths time.

C. Provide sets of cards 0–20. Let children stir their set around in a box, then see who can be first to order their set. This is best done working in pairs to reinforce language.

Whole class review

Time: 10 minutes

Key idea	Recognise the number pattern 21, 31, 41, …

- *'The group working with me will tell you about their number patterns.'*
- *'Tell us about the pattern counting in tens from 15.'* (25, 35, 45 etc)

Alternative starters

1. Play CG 16 Washing lines using numbers between 10 and 20.
2. Write some numbers beyond 10. Children match them to the 100 square and try to name them.

Lesson 2: In your head

Level: 1–2

Objectives:
- adding 1 mentally and with cubes/fingers
- recognise and name numbers to 100

Whole class starter

Time: 10 minutes

1. Recite numbers with the children as far as most are reasonably confident. Point to the numbers on the 100 square. Ensure that children understand that a 100 square is a squashed up number line by doing MM 6.
2. Play CG 26 In your head, giving children plenty of practical experience of the language of addition (and subtraction; teaching the language of addition and subtraction together helps make connections).
3. Point to a low number, eg 3. *'Show me that many fingers.'* Focus on adding 1. Ask simple questions, eg *'What number comes next?' 'What is one more than 3?' 'Shut your eyes and pretend you are standing on 3 on the number line. Take one step. What number have you landed on?'*
4. Give several examples. *'Dan, come and point to 5. Now hop on 1. What is 1 hop after 5?' 'What if you take one hop back again? Yes, you end where you started.'*
5. Demonstrate the pattern of adding 1. *'3 add 1 is 4. 4 add 1 is 5.'* Circle the numbers. Children use fingers or cubes.
6. Practise mental maths, including subtraction. *'Shut your eyes. Think of 3. What number is one less than 3?' '3 bears and 1 goes home. How many are left?'*

Group time

Teacher focus

Time: 15 minutes

1. **Support:** Work with those children who you know need help with PB 3 p 8. Help them to see the patterns made by adding.

Working more independently

A. PB 3 p 8

B. **Maths table:** 2 different colour pens, 2 counters, dice (1–6 or 4–9 or to suit your children). Lay the IP down so that 2 children can play a race game on the 100 square. (It would be very slow with more than 2.) Take turns to throw the dice and move your counter that many spaces. Draw a circle on numbers you land on in your colour pen. Race to 100. Let children start at 20 if time is short.

C. Use cards 0–10 and + and =. Children pretend to be a teacher and make sums for the rest of the class at review time.

Whole class review

Time: 10 minutes

Key idea	Adding 1 or 2 to a number.

- Read the numbers circled on the maths table game. Ask *'What is 1 more/less than that number?'*
- Do some 'in your head' mental maths, eg *'Work out 2 and 1 more, 6 and 1 more, 1 less than 9.'*

Alternative starters

1. Pair the children putting a more able child with a less able. Using cards 1–20 or 50/100, deal out some random cards to each pair. Children lay these down in front of them. Start at 1, let children offer their cards as they are needed. Keep the cards in order for the next session.
2. Give small groups cards, 10–19, 20–29 etc. They muddle their cards (on a table to keep their set together). Groups race to order their cards.
3. Cover several numbers on the 100 square. *'Which numbers are hidden?'* Let children write or tell you about the numbers. ('A three then a four' is easier for some children than 34 and will help children to see patterns.)
4. Make a large blank 100 square. Ask children to tell you what to write to fill in 1–100. This can be completed as a maths table activity.

Lesson 3: Number patterns

Level: 2

Objectives:
- counting in 2s and 10s
- number patterns

Time: 15 minutes

1. Read 'Number patterns' *Seven Dizzy Dragons*, p24. Circle the numbers on the 100 square and relate these to a number track. Say:

   'I can count, way up high
   count up in steps of 10,
   10, 20, 30, 40 ...'
   (Let children chant the number pattern.)
2. Relate this to counting in 10s to 100 by using 10 'ten strips' of cubes. Read 'The chimpanzees' tea party' *Seven Dizzy Dragons*, p12.
3. Count in 2s, circling the pattern of the numbers. Read 'Noah's ark' *Seven Dizzy Dragons*, p11. Chant 2, 4, 6, 8, etc.

You need:
ten-strips, objects to count number cards, playdough

Key words:
ten, twenty, thirty, ... hundred, every other number, see a pattern, straight line, diagonal line

Working more independently

A. **Maths table:** Have a Noah's ark and animals for counting in 2s, and number cards 10, 20, 30, ... to count in 10s.
B. Make 10 playdough chimps (or use 10 soft toys) and count out cubes for some of their food.

IP page 16 **Picnic**

focuses on sharing both unequally and equally by separating a number of objects into 2 or more groups.

Learning objectives:

- sharing unequally and then equally between 2, 3 and 4
- language of lots of and sharing
- counting in 2s and 10s

Lesson 1: Dragon is hungry

Level: 1–2

Objective:
- splitting up a number of objects into 2 unequal groups

Whole class starter

Time: 15 minutes

1. Cover the left side of the page so that just dragon and ted can be seen.
2. Play CG 30 Dragon is hungry.
3. On the IP, draw a number of buns, say 7, on the table cloth (or use Blu-Tack to stick on paper buns) and write the numeral.
4. Ask children to come up and move the buns or draw their suggestions for splitting the buns. '*Show us one way you could share the 7 buns unequally so that dragon has more.* (eg dragon 6 and ted 1) *How many more does dragon have? What if we give ted one of dragon's buns. How many more does dragon have now?*'

You need:

2 paper plates, buns (cubes), paper buns, Blu-Tack

Key words:

more/fewer than, unequally, equal, compare, left over

Initial assessments:

Note who

- can split up numbers using appropriate language.

Group time

Time: 15 minutes

A. Move the children into groups, repeating the starter activity but with groups differentiated by the numbers that they use.
'Split up 8 buns unequally so that hungry dragon has more than ted and then draw what you did.'

B. **Maths table**: *'Make a meal for Jack and the giant with 10 sandwiches and 10 cups of juice. Give the giant lots more than Jack so that the giant won't be hungry and eat Jack! Record what you do in your own way.'*

Whole class review

Time: 10 minutes

Key idea	Numbers can be split in many different ways.

- Compare and talk about drawings, emphasising the language of sharing and comparison.
  'Tell me a way that I could split up 10 buns?'
  'Is there a different way?'
  'How many more is … than …?'
  'Is 3 less than 7 or more than 7?'
  'Tell me 2 other numbers. Which one is more than the other?'
  'Jake, count out 3 buns. Shirley, count out 4. Who has the most buns?'
  'On a number line, starting from zero, would you get to 3 or 4 first?' 'How many steps from 3 to 4?'

Lesson 2: That's not fair

Level: 1–2
Objective: ● recording unequal sharing

Introduce the lesson using the bottom border of the IP. Choose numbers and show how to record for RS 9.

- Read 'Chook, chook' *Seven Dizzy Dragons*, p 23.

Lesson 3: Make it fair

Level: 1–2
Objective: ● splitting up a number into equal groups

Whole class starter

Time: 10 minutes

1. Cover 2 of the characters.
2. Do a simple demonstration with two children with 6 buns/apples and ask children to tell you how to share out the apples equally. Be explicit that this is different from unequal sharing.
   'How is it different?'
   'Which other number shall we try to share equally between 2?' (Take all suggestions and try them out. Reassure children who suggest an odd number: *'Oh dear, look what is happening. What is special about having 7 buns? Let's make a note of 7 and see if there are any more numbers like it that won't share out equally.'* Let children use fingers to explore numbers.

Group time

Teacher focus

Time: 15 minutes

1. Introduce this activity to the whole class. Put children into groups of 2, 3 or 4 with that many plates and related number cards, eg 3, 6 and 9 for groups of 3. Ask children to count out as many buns as one of their number cards, then share them out equally. Ask them to record what they do in their own ways 'so that you will remember what happened' and to bring it to the review session.
2. **Extension**: Give out cards 0–20. *'Which numbers share equally between 2 groups* (and 3 and 4 groups if they are able to do that)?'

Working more independently

A. Maths table: Use cups and plates from the role play area to share out playdough food equally.

You need:

cubes, teddies or buns, paper circles or plates, number cards in sets as follows:
- even numbers that go into 2 groups equally.
- 3, 6, 9, and 12 that go into 3 groups equally.
- 4, 8, 12, 16 that go into 4 groups equally.
- Extension: a set of number cards 0–20.

Key words:

sharing equally so that it is fair

Initial assessments:

Note who
- volunteers information about numbers that will share equally between 2.
- is sharing equally.

Whole class review

Time: about 10 minutes

Key idea	Sharing out fairly.

- Compare and talk about drawings made during group time, emphasising the language of sharing equally and how each person has a fair share in equal grouping.
- Be explicit that this is different from unequal sharing and when they split up numbers or things such as buns, they must know whether they are going to do it equally or unequally.
- *'Tell me which numbers work well when sharing equally between 2?'*
- *'Can you share out 10 buns between 3 people?'* (No, but some children might well say they could cut one bun into 3 bits!)

Lesson 4: Lots of

- language of lots of and sharing

Repeat sharing and use children's drawings to talk
about 'lots of'.
*'Sheryl had 12 buns and she shared them with Anna so
they had 6 buns each. One lot of 6 on this plate, another
lot of 6 on the other plate. How many lots of 6?'*
'Who can tell me about "lots of" on their picture?'

Number track

1	2	3	4	5	6	
7	8	9	10			

First Skills in Numeracy 1 © Cambridge University Press 1998

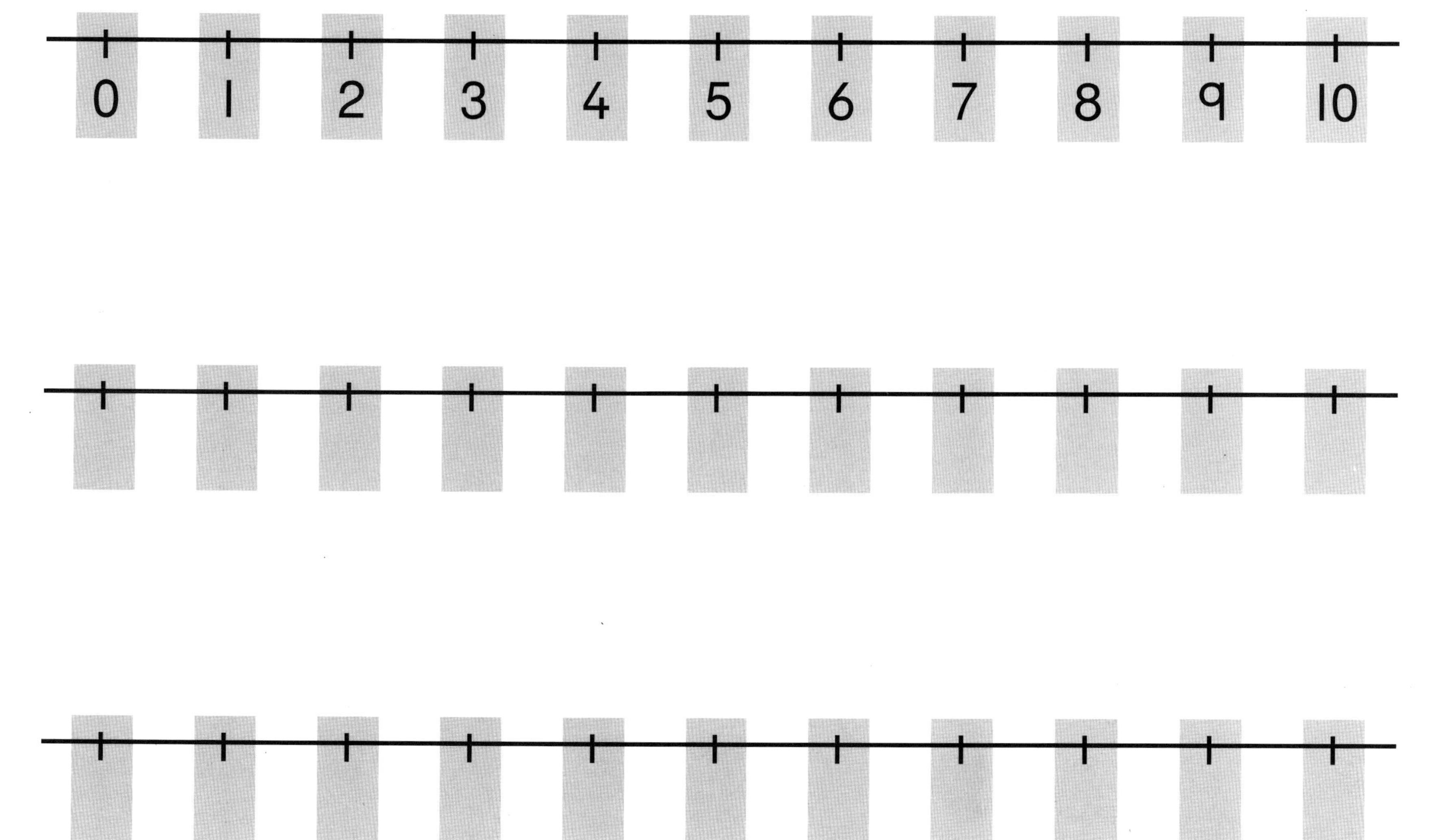

Name .... Date ....

Draw ● to make

Draw ● to make

Draw ●

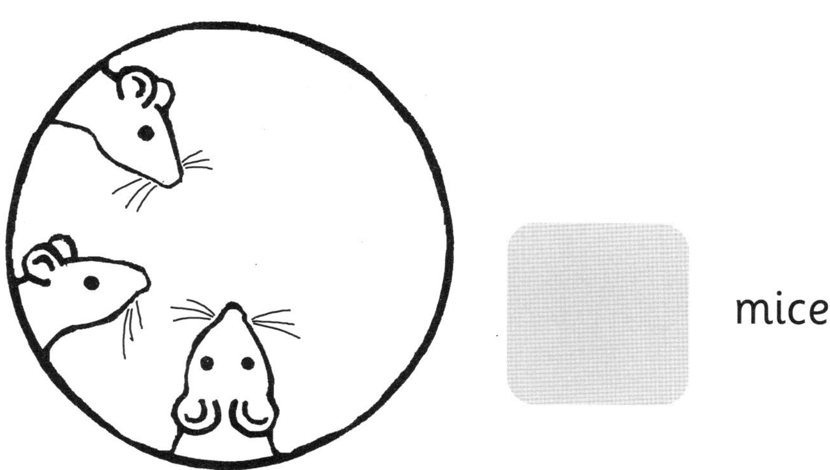

mice

Draw what you did

Name .... Date ....

Play CG29 'Splitting numbers' with your teacher.

I have  toys.

Draw what you did

First Skills in Numeracy 1   © Cambridge University Press 1998

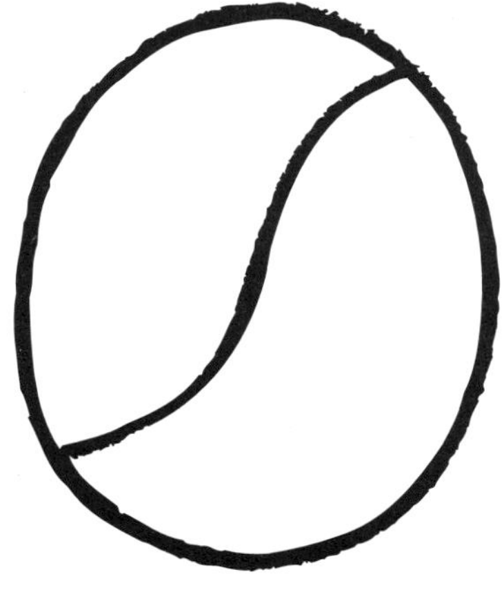  and  is 

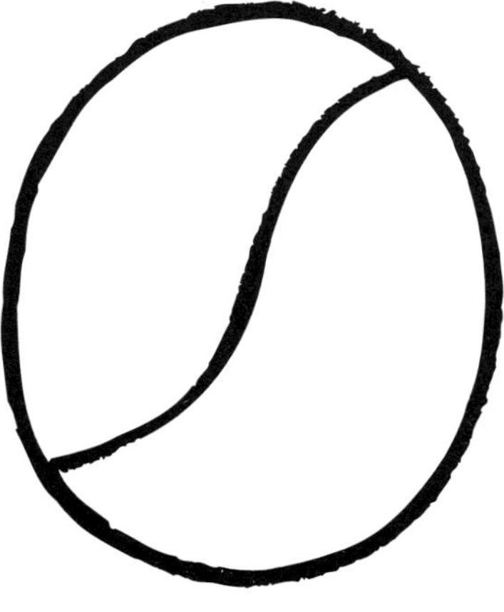  and  is

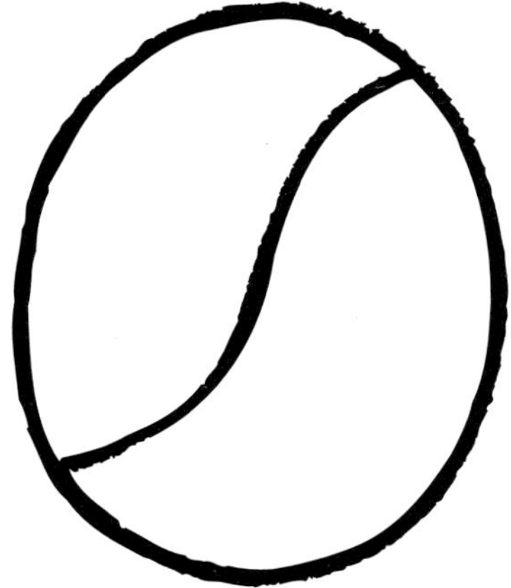

Start with ☐ children

Name .... Date ....

Split up numbers

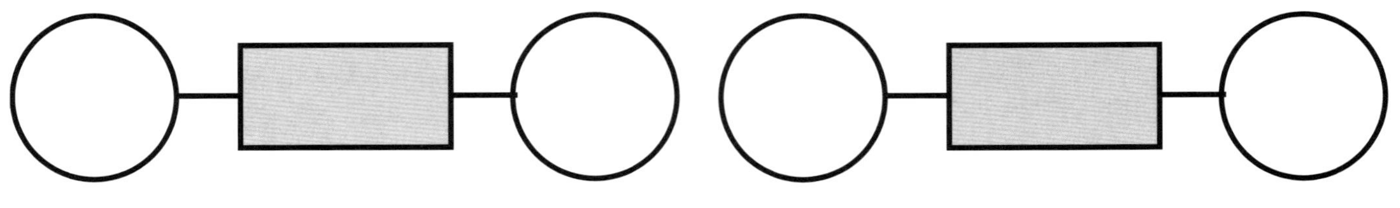

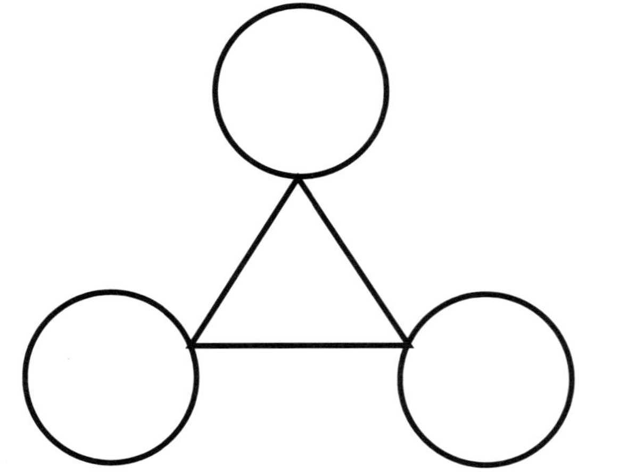

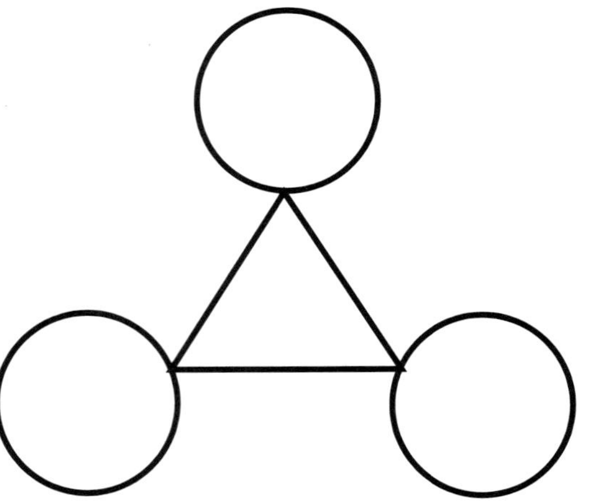

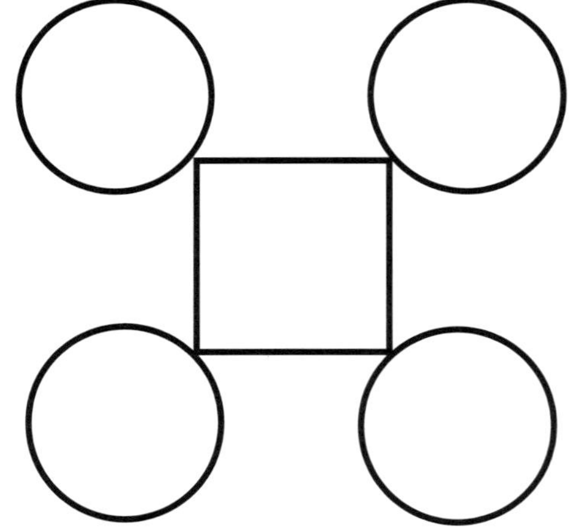

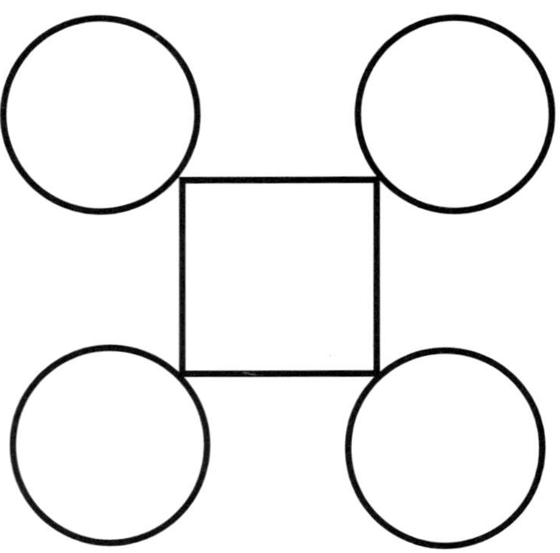

Choose

Planning grids

Use the grids to identify lesson plans and activities that develop the learning objectives that you want to teach.

There are two grids: one for nursery and one for reception. You may need to use elements from each to suit the differing abilities of children in your class.

Many of the learning objectives can be met in several activities; only the major learning objectives of each activity are listed.

Abbreviations:

MM Mental maths activity; pp 17–19
CG Circle game; pp 20–37
P Structured play activity; pp 39–42
MT Maths table activity; pp 44–7

Nursery objectives and First skills 1, links

Summary of objective	Nursery lesson plan	Activity bank
Early data handling, matching, sorting, comparing		
● Making 'go togethers', 1 criterion teacher led, progressing to 2		MT 2
● Making 'go togethers', any consistent criteria		CG 2
● Sort objects with mathematical criteria (shape, size, ...)	9b	MM 2, 3; CG 1; MT1
● Early language of comparing, same as, nicer than, ...	9b	
● Match, order, compare with mathematical language, eg more than, larger than	9a	
● Collect, discuss data		P 5
Counting		
● Recite first to 3/5 then above	1b, 1c	CG 3–5
● Count and match, eg numbers and dots 1, 2	2c, 2g	
● Early counting groups of objects to 3/4/5	1a, 2a, 3b, 4a, 5b, 5c, 6b, 6c, 7a, 7b, 8a, 8b, 11a, 15a	CG 8, 10; P5; MT 5–9
● Counting, one step one name (one-to-one correspondence)	2f, 5a, 8f, 8g	
● Knows last number in the count is the number in the set	1c, 7a	
● Conservation of number	1a, 1b, 2a, 5a	CG 11
● Talk about familiar large numbers	6d	CG 7
● Recognise familiar layouts of numbers / dice patterns	2c	

Summary of objective	Nursery lesson plan	Activity bank
Representing, reading and writing numbers		
● Recognise numbers / distinguish from letters	2b, 8c	
● Represent numbers in own way		P 5, 8; MT 5–7
● Tally / keep score in own way, one mark each number		P 5
● Match numerals/fingers to numbers of objects	1a, 2c, 4a, 8g	CG 8
● Read/write numbers to 3/5/10	1c, 5b, 7a, 8d, 9c, 11b, 12a	CG 13, 14; P 6
● Begin to read larger numbers	14b, 15d	
● Begin to read number names		CG 13, 14
● Visualise and talk about numbers		MM 5, 6
Comparing and ordering numbers		
● Begin to understand and use the vocabulary of comparing and ordering numbers	4a, 6a, 6b, 6c, 11a	P 6; MT 3, 4
● Compare two familiar numbers and say which is more or less	3b	MM 11
● Order a complete sequence of numbers	2b, 8c, 15b	MM4; CG 16–18
● Begin to understand and use ordinal numbers	6a, 8e	
● Early number line experience	8f	
● Take steps of one along number track	14a	
● Language of one more / move on one / how many steps to get to …	8f, 14a	
● Count up and down the number track	8f, 14a	
Calculations		
● Know / work out with fingers simple number bonds	11c	
● Early experience with addition and subtraction	7b, 11c	MM 7
● Begin to understand and use vocabulary of addition and subtraction		MM 8, 9
● Add or subtract one to/from a number		MM 10, 11
● Begin to understand addition as counting on	7b	
● Separate a number of objects into two groups		CG 28; MT 13
● Remove a smaller number from a larger and count the remainder		CG 33
● Remove a smaller number from a larger and find the remainder by counting on/back		CG 37
● Separate a number of objects into three groups		MT 13
● Mental addition/subtraction	10a, 10b	MM 14
● Begin to know number bonds to 10		MM 12, 13; CG 28, 35

Summary of objective	Nursery lesson plan	Activity bank
Making sense of number problems		
Problems involving 'real life' or money		
• Solve simple 'real life' problems in a practical context, deciding what to do, eg sharing	1d, 16a	
• Use the language of simple problems / how many altogether?	Throughout, especially play	
• Predict outcome of simple problems, add / take away one more?	8f	
• Begin to recognise coins and use them in role play situations to find totals and give change		P 4, 5
• Arrange objects in pairs		CG 2, 14; MT 2
Problems involving measures		
• Language: tall, medium, short, wider, thinner, …	2d, 2e, 3a, 4b, 12b	CG 40–2; P 1, 2; MT 15, 16
• Language: full, empty, holds more/less, …		P 1
• Language: heavy/light/weighs more than, …		CG 40; P 1, 3; MT 17
• Language: fast/slow	4b	
• Begin to understand and use the vocabulary related to time; sequence events; read o'clock times	3c	
Shape and space		
• Begin to recognise solid and flat shapes and talk about properties such as whether a solid shape rolls or slides or how many sides or faces it has	9a	MM 15; CG 43
• Recognise and recreate repeating patterns		CG 38; P 7, 8
• Begin to describe the position, direction or movement of an object using everyday words; respond in PE and other activities to instructions about position, direction or movement	3d	

Reception objectives and First skills 1, links

Practice books are provided for children who are ready to record their work more formally. PB pages are referred to in the reception lesson plans and learning objectives are given on each PB page.

PB1: deals with counting, reading and writing numbers to 6 and above.

PB2: extends the reading and writing of numbers to 10 and introduces addition.

PB3: develops addition and subtraction.

References in the second column are to the lesson plans on pp 62–127; for example, 8(2) refers to the reception lesson plan for Interactive picture 8, lesson 2,

Summary of objective	Reception lesson plan	Activity bank
Numbers and the number system		
Counting		
● Recite the number names in order from zero (first up to 10, then to 20, then beyond)		CG 3, 4
● Recite the number names in order, counting on from a given number	1(2)	CG 5, 6
● Recite the number names in order, counting back from a given number		CG 6
● Know some of the number names for larger numbers, and other numbers in their daily lives		CG 7
● Count reliably a collection of objects (first up to 10, then 20, then beyond), with one-to-one correspondence (giving just one number name to each object)	1(1, 2), 2(1, 3), 5(1), 8(3), 9(1), 11(1), 14(1), 15(1)	CG 8, 9; P 5; MT 2, 5–9
● Count reliably in more difficult contexts, such as sounds or movements		CG 10
● Conservation of number		CG 11
● Count in tens	14(3), 15(3)	CG 12
● Count in twos	1(3), 15(3)	CG 12
● Visualise numbers as 5 and a bit	14(1)	
Reading and writing numbers		
● Begin to read numbers and number names	2(1, 3, 4), 5(1), 8(3), 15(1)	CG 13, 14; P 6
● Begin to record then write numbers, initially using marks or drawing 'beads' progressing to tallying and writing symbols	4(2), 5(1), 8(4)	P 5, 6, 8; MT 5–7

Summary of objective	Reception lesson plan	Activity bank
Numbers and the number system		
Comparing and ordering numbers		
• Begin to understand and use the vocabulary of comparing and ordering numbers	1(2), 4(1, 2), 5(2), 11(1, 2)	MM 4; CG 15; MT 3, 4
• Compare two familiar numbers and say which is more or less	2(4)	MM 11
• Order a complete sequence of numbers	8(1, 3), 9(1), 14(2), 15(1)	CG 16, 17
• Say a number lying between two others		CG 18
• Order some selected numbers		CG 19
• Begin to understand and use ordinal numbers	6(3), bottom of IP 8	
Estimating numbers		
• Estimate a number in the range that can be counted reliably, and check by counting		CG 20
• Estimate by choosing the more likely of two given numbers, including numbers beyond their counting range		CG 20
Fractions		
• Begin to recognise and use halves in context		CG 21; MT 14

Summary of objective	Reception lesson plan	Activity bank
Calculations		
Understanding addition and subtraction		
Working with a familiar range of numbers in practical contexts:		
• Begin to understand and use the vocabulary of addition and subtraction	10(1), 11(1–4), 13(3), 14(2)	MM 7, 9; CG 22; MT 10
• Add or subtract one to a number	11(2), 13(1, 2), 15(2)	MM 10; MT 11
• Begin to understand addition as the combination of two groups of objects; extend to three groups of objects	2(5), 6(1), 13(3, 4)	MM 8; CG 23–5
• Begin to understand addition as counting on	8(1, 2), 10(2), 14(1, 2)	CG 26
• Begin to recognise the addition of doubles by counting on		CG 27; MT 8
• Separate a number of objects into two groups	5(4), 7(2), 12(1), 13(1, 2, 5), 16(1–3)	CG 28–31; MT 13
• Combine two groups to make a particular number of objects	5(4)	CG 28–31
• Remove a smaller number from a larger and count the remainder	7(1), 11(4)	CG 33, 34
• Remove a smaller number from a larger and find the remainder by counting back from the larger number	10(2), 14(2)	
• Begin to find out how many have been removed from a larger group of objects, either by counting back to a number, or by counting up from a number	8(1, 2)	CG 35
• Work out how many are needed to make a larger number		MM 12, 13; CG 32, 36
• Begin to find the difference between two numbers, either by counting back or by counting up		CG 37
• Separate a number of objects into three groups	12(2), 16(3)	MT 13

Summary of objective	Reception lesson plan	Activity bank
Making sense of number problems		
Reasoning about numbers		
● Make, describe and continue simple patterns	6(3), 15(3)	CG 38
● Solve simple number puzzles in a practical context and respond to: 'What could we try next?'		CG 39
● Make predictions		CG 20
Problems involving 'real life' or money		
● Solve simple 'real life' problems in a practical context, deciding what to do	Throughout, especially play	
● Begin to recognise coins and use them in role play situations to find totals and give change	5(3), 6(2)	P 4
Problems involving measures		
● Begin to understand and use the vocabulary related to length, mass and capacity	3(1, 2)	CG 40–42; P 3
● Make direct comparisons of two lengths or masses, and fill and empty containers	3(1, 2)	P 1, 2; MT 15–17
Time		
● Begin to understand and use the vocabulary related to time; sequence familiar events; begin to know the days of the week and read o'clock time		P 3
Data handling		
● Sort objects, pictures or themselves, according to one criterion, progressing to two criteria		P 5

Summary of objective	Reception lesson plan	Activity bank
Shape and space		
Vocabulary		
Begin to understand and use in practical contexts:		
• names of shapes such as cube, cone, circle, triangle, square, rectangle…	9(2)	
• flat, curved, round, straight, solid, hollow, corner, face, edge, side…	9(2)	
• symmetry, symmetrical…		CG 44
• position, above, below, over, under, top, bottom, on, in, outside, inside, in front, behind, opposite, around, beside, next to, between, middle…	9(3)	CG 45
• direction, left, right, up, down, forward, backward, across, along, through, towards, away from…	3(4)	
• movement, roll, turn, slide, stretch	3(4)	
3D and 2D shapes and their properties		
• Begin to recognise solid and flat shapes and talk about properties such as whether a solid shape rolls or slides or how many sides or faces it has		MM 15; CG 43, 46
• Make and talk about shapes and patterns using construction kits, bricks, plasticine, play dough, cutting out, stamping, printing…		P 7, 8
• Talk about symmetry in leaves, insects, people, buildings, patterns…		CG 44; P 7, 8
Position, direction and movement		
• Begin to describe the position, direction or movement of an object using everyday words; respond in PE and other activities to instructions about position, direction or movement	3(4)	CG 45